The Downloader's Handbook

Your complete guide to using broadband for downloading, ripping and converting music and film

by David Stevenson

HARRIMAN HOUSE LTD

43 Chapel Street
Petersfield
Hampshire
GU32 3DY
GREAT BRITAIN

Tel: +44 (0)1730 233870
Fax: +44 (0)1730 233880
Email: enquiries@harriman-house.com
Website: www.harriman-house.com

First published in Great Britain in 2006 by Harriman House.

Copyright Harriman House Ltd

The right of David Stevenson to be identified as the author has been asserted in accordance with the Copyright, Design and Patents Act 1988.

ISBN 1-897-597-65-7
978-1-897597-65-1

British Library Cataloguing in Publication Data
A CIP catalogue record for this book can be obtained from the British Library.

All rights reserved; no part of this publication may be reproduced, stored in a retrieval system, or transmitted in any form or by any means, electronic, mechanical, photocopying, recording, or otherwise without the prior written permission of the Publisher. This book may not be lent, resold, hired out or otherwise disposed of by way of trade in any form of binding or cover other than that in which it is published without the prior written consent of the Publisher.

Printed and bound by Graficas Cems, Navarra, Spain

No responsibility for loss occasioned to any person or corporate body acting or refraining to act as a result of reading material in this book can be accepted by the Publisher, by the Author, or by the employer of the Author.

Information has been obtained by David Stevenson from sources believed to be reliable. However, because of the possibility of human or mechanical error by our sources, David Stevenson does not guarantee the accuracy, adequacy, or completeness of any information and is not responsible for any errors or omissions or the results obtained from the use of such information. All information was believed to be correct at the time of writing.

Designated trademarks and brands are the property of their respective owners.
"Microsoft, Encarta, MSN, and Windows are either registered trademarks or trademarks of Microsoft Corporation in the United States and/or other countries."

Every effort has been made to trace and contact all copyright holders, but if any have been inadvertently overlooked the publishers will be pleased to make the necessary credits at the first opportunity.

Contents

Preface	vii
Introduction	xiii
1. Computer Hardware	1
Thirteen upgrade steps	5
Step 1 – Memory	5
Step 2 – USB and Firewire slots	9
Step 3 – Hard disk storage	11
Step 4 – Motherboard	16
Step 5 – CPU	17
Step 6 – Graphics card	20
Step 7 – TV tuner card or device	23
Step 8 – Operating system	24
Step 9 – Sound card	25
Step 10 – Home network	27
Step 11 – Streaming	30
Step 12 – Modem and broadband	31
Step 13 – Monitor	34
2. Securing Your Computer	37
Five steps to safe downloading	40
Step 1 – Update your software	40
Step 2 – Know the enemy	42
Step 3 – Anti-virus software	45
Step 4 – Spyware	54
Step 5 – Firewall	67
3. Digital Music	83
Ripping software	86
Welcome to the world of Nero	86
The Swiss army knife alternative – dBpowerAMP	89
Codecs galore – which ones to use?	91
How do they compare?	94

Digital music players	96
Windows Media Player	96
jetAudio	107
Winamp	112
Portable MP3 players	116
Internet radio and streaming audio	119
Live365	120
BBC	121
Recording music and speech using your PC	123
Audio Recorder Pro	123
The Audacity alternative	127
4. Online Music Services	**131**
The world of DRM	134
The major music download services	137
MSN	137
Apple iTunes	145
Napster	162
eMusic	168
Audible	174
Other services	177
Which service to use?	179
Range of music	179
Cost	180
5. Digital Video	**183**
Video streaming	187
Websites with great film content	190
Playing video files on your PC	191
The Codec War – DivX vs XviD	195
Compressing video – the DivX/XviD revolution	197
Working with DivX and XviD	198
DivX Create Bundle	198

More weird file types	202
RAR files	202
Cue and Bin files	205
Playing video files in the living room	207
Backing-up your dvds	209
Some DVD basics	210
6. File Sharing Networks	**239**
Content galore, pity about the law	241
What exactly is a file sharing network?	242
Which networks to use	245
BitTorrent – the new kid on the block	249
eDonkey/eMule	261
Other networks	268
Morpheus	269
BearShare	273
LimeWire	276
7. Copyright And The Law	**279**
Questions about copyright	282
Conclusion	**305**
The not too distant future – what to watch out for	307
Index	**314**

Preface

Who this book is for

If you want to download music and video files from the internet and listen to them on your PC, this book is for you.

The book is not written for techies, although the chapters on file sharing networks and copyright should be interesting for them. Rather, this book is for the ordinary PC user. Someone, say, who knows how to use a PC for writing letters and email, but has not played about much with music and video on the computer.

Finally, this book is for PC, not Mac, users. On Planet Mac things are very different – and arguably better organised and easier – but this book isn't for you.

How the book is structured

This book takes you by the hand and walks you through every step required to start using your computer as an entertainment centre.

I'll quickly run through the chapters in this book.

1. Computer Hardware

The books starts off with a checklist of minimum requirements your computer should have to play music and video. This includes things like the memory, hard disk, sound card and monitor. I also explain how to upgrade each component, if this proves necessary.

2. Securing Your Computer

Your next step is to make sure your computer is safe and secure. That's the subject of this chapter. I'll talk you through all the security basics you need to make sure you're safe online. At the end, I'll run you through a ten point checklist of 'Safety Essentials'.

3. Digital Music

In this chapter I look at the various software-based music players available, and the various ways that music from a CD is turned into a digital format and then compressed into files small enough to store by the hundred.

After you've read this chapter you'll be able to:

1. 'Rip' a CD onto your hard drive.
2. Play that music on your PC.
3. Convert that music into a smaller file that can be played on any PC or portable MP3 player.
4. Learn about different file formats, including MP3, MP4, WMA, OGG and MPC.
5. Work out how to best use the popular Windows Media Player.
6. Record internet radio broadcasts.
7. Decide which portable MP3 player to buy.

4. Online Music Services

We then move on to online music services. We go online and sample some of the fantastic, legal music download services like iTunes and Napster. I'll also introduce you to some smaller, but equally brilliant, music services like emusic and Wippit.

In this chapter you'll learn how to:

1. Use the iTunes music player and download music from the iTunes music store.
2. Sign on and use the Napster service.
3. Access the superb emusic online service with its wealth of alternative music.
4. Work with the Audible audio books service.

5. Digital Video

In the digital video chapter we look at the fastest growing area of internet downloading – film and video. More and more users are moving beyond music and starting to download video such as films and pop promos. But the technology involved with these downloads can, initially, be quite daunting. The world of video is full of nasty sounding formats and even stranger sounding codecs. And if you ever decide to get into the business of backing up your DVD films onto a hard drive (a great idea for all you film fans) you could soon find yourself sinking into an ocean of strange terminology and even stranger programs.

Don't panic! This chapter will teach you how to:

1. Stream pop promos and short movie clips over the internet.
2. Work with codecs that turn movies into relatively small 600MB files.
3. Use simple media players that will play virtually every format under the sun.
4. Back up your DVDs onto your hard drive.
5. Get compressed movies onto a DVD that can then be played on your TV.

6. File Sharing Networks

Once you've acquainted yourself with all the codecs and players, it's time to start sourcing the content online. That's where the chapter on file sharing networks comes in – it'll walk you through the process of deciding which network to use and then explain what these networks consist of and how to use them. After reading this chapter you'll understand:

1. The principle behind file sharing network;
2. How to work out which network to use;
3. The wonders of BitTorrent; and
4. How to use the slightly more complicated eMule network.

7. Copyright And The Law

In the last chapter we get to the knotty issue of copyright – what's legal, and what isn't.

How this book will help you

By the end of this book you should be able to:

1. Play all types of music on your computer.
2. Rip and burn music.
3. Work out what kind of MP3 player to buy.
4. Download music from any of the major online music services.
5. Work with every major video film format.
6. Compress a film file into 600MB
7. Back-up all your DVDs on to your computer.
8. Download tens of thousands of different files available on BitTorrent and eMule.
9. Know how to make your computer safe and secure.
10. Understand your legal position.

How to read this book

For some readers, especially those whose computer skills are rudimentary, I'd recommend going through this book reading each chapter in turn. More experienced computer users can use this book as a reference guide, dipping into chapters where they want to improve their knowledge.

I provide a rough guide below to how certain types of readers might want to approach this book.

- **Patient beginners**
  These readers should start with chapter one, and then work step by step through the whole book. The first two chapters may not be that exciting, but they are extremely important, especially the second chapter on security. The internet offers an incredible opportunity to download music and video, but there is a cost, and that cost is extra vigilance. Charge onto the internet, downloading without making sure your computer is secure, and I can assure you grief will follow!

- **Impatient beginners**

    If you really want to get straight to the meat, then by all means jump straight to chapter 3 on digital music, or chapter 5 on video. This will give you a flavour of what's possible. But then, before launching yourself into downloading madly, do read chapter 2 to check that your system is secure.

- **Advanced**

    For advanced readers, you might like to skim the early chapters, and then go straight to those on file sharing networks and copyright.

Remember computing is a fast moving industry. Every effort has been made to ensure that all the information was as accurate and up to date as possible at the time of writing.

Supporting website

The website supporting this book can be found at:

www.harriman-house.com/downloaders

Introduction

The revolution has begun

My first memory of a computer is in the late seventies, when my dad brought home a very early Apple computer. I can't remember what my reaction was, but I do remember that it was very bulky, made of horrible plastic and seemed to do a small number of things, very slowly. After an hour or two playing around on it I decided it was about as exciting as a wet fish, and so I went back to kicking a football in the garden.

Many years passed before I finally got a proper PC – back in the mid nineties – and my enthusiasm was starting to smoulder into a passion. I'm no computer programmer or hacker – don't expect pages and pages of complicated code and programming language in this book – but I love computers for what they can do to make my life easier and fun.

And then the internet arrived, with its email and web pages. A fantastic development. But computers themselves, though a lot more powerful than my dad's early Apple, weren't exactly exciting. Just useful.

Fast forward ten years to today. Suddenly, ordinary home PCs are starting to do some amazing things.

Yes, computers are still used by most of us for mundane stuff like word processing or sending emails, but recently PCs have been transformed. They're now multimedia powerhouses.

Connect it to the internet and you can acquire music from around the world, edit it, and then build up huge compilations that can be played back on devices that have no resemblance to a PC. Video is now commonplace on computers as PCs turn into home entertainment centres. Everything enjoyable about our media-soaked world has started to migrate from the lounge – with its old-school electronics appliances – into the study, the home of the PC.

Let me give you an example.

I love films and music. In fact, I like them so much that I've acquired hundreds of DVDs along with thousands of CDs. But I hate CDs and DVDs, the stuff that these movies and music come on. I still remember that laughable *Tomorrow's*

World item that purported to show that CDs were virtually indestructible. The presenter smeared all sorts of sticky stuff on them and then tried to break them. Miraculously these sturdy CDs survived. Case proved.

Far from it. The *Tomorrow's World* report was, as we all know by now, complete cobblers. These early pioneers had obviously never scratched a CD or DVD; in my experience these optical media are about as reliable as vinyl, if not worse!

Step forward my media server.

I've collected my favourite films and music, taken them off discs via a technology called 'ripping', and then stored them on a single external hard drive that sits in my lounge. This disk now contains hundreds of albums and movies. Whenever I want to access this entertainment library I simply click on a file in my PC and I'm away – listening or watching the album or movie in any room of the house.

Call me a bit of a geek, but I love it.

Oh, and it gets better. I've added to this media library by downloading files from the internet. Music from Napster, TV programmes from the US and copies of internet radio broadcasts by my favourite DJs.

I present these small snapshots of my own personal computing experience to demonstrate a number of important points that underlie the thinking behind this book.

1. Computers are now fun and exciting. They involve stuff that excites and interests all of us – digital content like film and music. It's fun. It's not just data processing and programming.
2. With the growth of the internet, PCs have started talking to other computers; and as they've done this they've discovered even more compelling content online. Some of this is legal, some less so. But there's a world of content out there, waiting for you!
3. This book is written from a user's perspective. I don't come at this new world of networked, digital content from the perspective of the techie programmer trying to explain the impossible. I use all the stuff in this book, if not every day, pretty much every week. If I can use what's described in this book, so can you.

My message is simple: get hi-tech, get multimedia, get online and get broadband.

Why?

- It's worth it. The online content available is truly awesome.
- Everyone else is doing it. Broadband and internet usage rates in the UK are skyrocketing – in fact they're skyrocketing globally. And most of that internet traffic is in digital content like film and music.
- It's easy. This book will guide you through everything you need to know to access this new world of digital content, both offline on your PC and online through the internet.

Ten statistics to mull over

Let me shoot ten random statistics at you that underline the scale of this multimedia revolution.

1. The legal music download service, iTunes, boasts 1.7 million songs online; the equivalent of more than 100,000 albums. Rival Napster has 'only' 1.6 million.
2. Online music service Napster has over 750,000 registered users in the UK alone. That's 750,000 more than two years ago.
3. 60% of all traffic on the internet is currently originating from a bunch of file sharing networks with silly sounding names like BitTorrent, eDonkey and Morpheus.
4. 60% of all that traffic is now in film, not in MP3 music tracks.
5. As of August 2005, there are estimated to be just under 10 million global users online using file sharing networks at any one time.
6. At one internet service provider, 30% of all its traffic one day was just one file (a film) being shared over multiple networks.
7. Online music networks and file sharing networks now boast files that total 10 petabytes of data. I'm not going to specify how big that is, except to say that it's thousands of times the amount that existed on Planet Earth just a few decades ago.
8. An estimated 35 million Europeans regularly admit to downloading music and film from legal and non legal networks over the internet.
9. Just one year ago, the majority of UK internet users were using dial-up networking to access the internet. That meant they surfed the internet at maximum speeds of 56Kbps. By summer 2005, the number of broadband users had exceeded that of dial-up users – all from a standing start just four years ago. That means broadband users can now surf the net at speeds of at least ten times those of dial-up users. Most speeds are 40 times this level and some cable networks offer speeds that are 80 times dial-up speeds.
10. The availability of broadband is expected to reach 99% of the entire UK population within a few years.

Introduction

I'm not expecting you to understand absolutely everything in these statistical snapshots, but hopefully you'll get the idea there's a revolution out there!

If this revolution had a chief ideologue (think Lenin for the Russian Revolution) I feel confident he'd make the following statements:

- Nearly all music and film content has now been turned into digital content and is available in some form or shape online. At the core of this revolution is something called ripping – turning analogue film and music into digital files that can be stored on your PC.
- This huge mountain of content is available now for download online through either legal, or less than legal, networks.
- These networks have been designed to be easy to use, especially if you have a computer that can now double as a multimedia machine. Your computer has turned into tomorrow's entertainment centre.

Millions of users globally are online, downloading content day in, day out. You can either ignore the revolution or start downloading now!

Computer Hardware

Before getting started on this chapter, I'd repeat what was said in the preface, which is that those readers impatient to get at the sexy music and video content, might think about skimming this chapter and the next, and go straight to the third chapter on digital music.

Computers constantly need upgrading.

In fact, I'm willing to bet that every year you'll need to buy either new hard disk space or extra memory. It's just an annoying fact of (computer) life.

Every few years you'll also find that you need to buy either a new monitor or even a new motherboard. As more and more digital content – music, film, photos – piles up on your hard drive, the technology you'll need to process it will need to change.

In this chapter we'll run through the essential upgrades – and new additions – you'll need for your computer if you want to get serious about downloading multimedia content. But first it's essential that you make sure your current computer is up to a basic spec.

Here's my current recommended minimum

- A 2.6Mhz CPU running Windows XP.
- 512MB of memory and 200GB of hard disk space (internally or externally).
- A 17 inch monitor, preferably 19 inch and for most users a flat screen display.
- At least 4 USB 2.0 ports and 2 Firewire slots.
- A CD/DVD burner, the minimum burn speed should be 4x if not 16x.
- Broadband access, with a minimum speed of 1Mbps and no capacity limits.
- A wireless network with broadband access.

If your current system is below that specified in the above box the bad news is that you'll need to upgrade fast. And, truth be told, even if your machine is bang up-to-date here's the really bad news: you'll need some key upgrades at the very least every two years.

In this chapter we'll run through the key upgrades in our thirteen upgrade steps, but here's a quick, very simple summary.

Upgrade summary

1. I confidently predict that each and every year you'll run out of **USB and Firewire** slots. Every couple of months you'll buy a whole new bunch of clever digital devices and they'll need to be connected to your PC. Get ready to buy some form of USB expansion regularly.
2. **Memory** - you can never have enough. You'll probably need to double your RAM every year or two. But don't worry about installing more memory - it's incredibly easy to do.
3. **Hard disk space:** you might choose to plumb in new hard drives inside your computer or buy an external drive. Whichever route you go you'll probably need to double your total capacity every two years.
4. It's a similar story with **graphics cards**. Want top range games graphics? Want to watch digital TV on your computer? You'll need a new card every two years!
5. DVD technology is moving so fast that you'll need a new **DVD burner** at least every three years.
6. Every three to four years you'll probably need to plug a new **CPU unit** in and probably upgrade your motherboard, although for most people a new computer might be the better option.

Thirteen upgrade steps

Here's my easy to understand thirteen steps to upgrade heaven!

1. Memory
2. USB and Firewire slots
3. Hard disk storage
4. Motherboard
5. CPU
6. Graphics card
7. TV tuner card or device
8. Operating system
9. Soundcard
10. Home network
11. Streaming
12. Modem and broadband
13. Monitor

We'll now look at each of these in more detail.

Step 1 – Memory

What is memory?

There are two types of memory associated with computers:

- *Memory*: short-term memory that is lost when the computer is switched off.
- *Storage*: long-term memory (e.g. hard disk) where data is retained when the computer is switched off.

In Step 1, we're interested in the first type of memory, that is memory that's used by operating systems and programs to temporarily store data while they are working. The bottom line? More powerful programs generally need more memory and that means the average ordinary home computer memory will usually take the form of RAM (an integrated circuit memory chip).

Why bother upgrading?

Increasing the amount of memory in a computer is one of the best computing investments you'll ever make. With greater memory, you can run more applications at one time, make them operate faster, and switch between them instantly. In general, most current operating systems run best with at least 128MB of RAM, and preferably 256MB or more to take advantage of the huge range of features the OS has to offer.

However, before you get too excited and rush out and buy lots of extra memory, you need to check out how many memory slots you have available on your motherboard – some limit you to two, while others provide up to four slots. Also you need to check if your computer requires SDRAM or the next generation – DDR – as the slots for each standard are very different.

Assuming your computer is reasonably modern, you'll have DDR slots (hopefully at least three of them) and the cost of upgrading should be minimal. Prices are coming down all the time but at the moment you can easily buy a good quality 512MB upgrade for about £35, while a 1GB memory upgrade will probably cost you between £70 and £80.

How much will I need and what make should I buy?

I've always bought brand name memory, usually from a big US company called Crucial (www.crucial.com). They're not the cheapest or necessarily the best, but their modules always work and they are affordable. They also have a great website that talks you through the upgrade process and tells you what you need.

Opposite is a table, courtesy of Crucial, with some recommendations for ideal memory size.

Table 1.1: Recommended memory

Typical use	Memory required
Simple computing Some word processing, occasional e-mail	256MB
Some business use Word processing and e-mail, spreadsheets, fax and communication business graphics, general gaming software, three or more applications open at once	384MB - 512MB
Heavy business use Word processing and e-mail, spreadsheets, fax and communication software, presentation software, illustration software, photo editing, Web browser	384MB - 512MB
Gamer and multimedia enthusiast Word processing and e-mail, photo editing, font packages and multimedia software, CAD software, CAM software, gaming	512MB - 1GB
Heavy Graphics Design 3-D CAD software, modelling software	2GB and up

Table 1.2: OS Software

Software title	Minimum requirements	Crucial recommendations
XP Pro	128MB	512MB
XP Home	128MB	512MB
Mac OSX	128MB	256MB
Windows 2000	128MB	512MB
Windows ME/98	64MB	256MB

Table 1.3: Design software

Software program	Minimum requirements	Crucial recommendations
Adobe Acrobat	64MB	128MB
Macromedia	256MB	512MB
Microsoft FrontPage	128MB	512MB
Adobe Illustrator	128MB	512MB
Adobe Photoshop	128MB	512MB
Adobe Premiere Pro video editing	256MB	1024MB +

Table 1.4: Games

Software title	Minimum requirements	Crucial recommendations
Half-Life 2	256MB	1024MB
Doom3	384MB	512MB +
The SIMS 2	256MB	512MB +
Star Wars Battlefront	256MB	512MB +

Is it a complicated upgrade?

Upgrading is easy. Total time including opening up the machine: 10 minutes.

Tip: When it comes to adding system memory, the general rule of thumb is: the more, the better. And with prices falling all the time it's not really worth buying much less than an extra 512MB. Also, do remember that most machines support only one type of RAM – with one type of module or connector – so mixing types isn't an option.

Note: The absolute minimum for heavy downloading and ripping is 512MB.

Step 2 – USB and Firewire slots

What are these slots?

The idea behind these input/output (I/O) ports is actually very simple. They can not only power up digital devices attached to them, but also allow these devices to communicate with your computer. Typically, you might connect printers and cameras to a USB (universal serial bus) port, and external hard drives (see section on storage) or video cameras to a Firewire slot.

USB

USB is the most established of the I/O formats and is standard on most machines made after the late 1990s. Older machines (pre-2002) will probably use the first standard (USB 1.1), while most modern machines will come with the newer standard 2.0. The only difference between these two formats is the speed at which the communications ports can handle data: USB 1.1 pushes data along at between 1 and 12 Mbps while USB 2.0 can handle bursts of up to 480 Mbps. In most other respects these two variations on the USB standard are the same – they can both be used to power some devices and they're both plug-and-play which means that most devices can simply be plugged in and Windows XP will recognize the device straight away. USB 2.0 slots are also particularly useful if you intend to add an external hard drive, as the data rate is more than fast enough to transfer music files across from the drive to your computer in a few seconds.

Firewire

Firewire (also known as IEEE-1394) is an innovative data transfer standard (devised by Apple) for moving big packets of data (typically film and video) between computers, very fast. On paper, its data transfer rate of 400 Mbps has been surpassed by USB 2.0, but talk to most video editors and they say they still use the original Firewire 400 standard. Firewire seems to be able to handle large blocks of big data files particularly well, which makes it the ideal slot for attaching hard drives to.

Why bother upgrading?

More and more of us are buying digital devices – phones, cameras, printers, scanners – that we want to connect to a PC. These devices all have a lead attached to them (although the wireless USB format promises to end this) that needs to be plugged into your computer. Traditionally, PC manufacturers built only 2 USB 1.1 slots onto the motherboard and, if you were lucky, a Firewire slot.

You need at least 4 if not 8 or more USB 2.0 slots attached to your computer, plus maybe another 4 Firewire slots. Think about it: you've probably got a USB printer, a scanner, maybe a couple of external hard drives, plus a camera, a video camera, and maybe even a phone that talks to your computer.

Upgrading your USB and Firewire slots

If you need to increase the slots, consider one of two options:

1. You can easily buy PCI slot cards that give you more ports, both of the USB 2.0 and Firewire variety. I'd buy one with at least 4 USB 2.0 slots and 2 Firewire.
2. The alternative is to daisywheel off the main USB slots and attach a bridging device – a USB hub for instance. These typically give you 4 USB slots and some even offer up to 7 USB slots.

If you don't have either USB or a Firewire (your machine will probably be more than four years old) but it does have a free PCI slot, you can add USB 2.0 and Firewire ports by slotting in a combined USB 2 and Firewire PCI card.

> **Tip**: Don't bother buying any I/O expansion slot that offers anything below the current USB 2.0. The new USB standard is much better and much faster – and there's hardly any price difference between USB 1.1 and 2.0. There's also a new Firewire standard about to hit the high street called Firewire 800 – it's a faster technology and should also soon become the standard, especially for video and film makers and those with big external storage drives.

Is it a complicated upgrade?

Like a memory upgrade, this is very easy. Simply open up the PC and find a series of white plastic slots in parallel with each other. All you do is insert the PCI card with the USB/Firewire slots on it into these slots and that's it – you've upgraded your ports to take more digital devices!

> **Note**: The absolute minimum for heavy downloading and ripping is 4 USB 2 ports and 2 Firewire ports.

Step 3 – Hard disk storage

What is it?

This is the hard disk drive in your system and it's the long-term memory data centre of your PC. It's here that all your programs and data are stored. But storage can also apply to other ways of storing data including flash drives, tape and removable drives, external hard disks, plus optical storage like CD-ROMs and DVD-RWs. Hard disk drives are almost as amazing as microprocessors in terms of the technology they use and how much progress they have made in terms of capacity, speed and price in the last 20 years. The first PC hard disks had a capacity of 10MB and a cost of over £50 per MB. Modern hard disks have capacities approaching 1,000GB and a cost of less than 20p per MB!

The amount of information a hard drive can store is measured in bytes. You'll need all the bytes, and megabytes, you can lay your hands on, as today's software programs (not to mention data, digital music, photos and multimedia) take up a lot of space.

> **Tip**: No matter what kind of storage technology you use, the best idea is to buy the one with the most storage capacity and the quickest access speed.

Lets start with the hard drive.

What do I need?

Most traditional hard drives built into a PC use a format called IDE (or Integrated Drive Electronics). IDE is an interface for mass storage devices in which the controller is integrated into the disk or CD-ROM drive.

Newer storage systems use an alternative standard called SATA (or Serial Advanced Technology Attachment). Serial ATAs are more expensive but:

- they offer faster data access rates; and
- they have smaller footprint leads.

You might also come across something called Raid, which allows data to be spread across different hard disk drives.

If you're going to store a couple of dozen films plus, say, a hundred albums you will absolutely need a minimum of 80GB in storage. To be honest, most heavy duty downloaders and eager beaver multimedia types wouldn't touch a hard drive under 200GB, and most of them are steadily switching over to SATA drives, which are only slightly more expensive than IDE drives.

Think about attaching external hard drives

External hard drives were traditionally the preserve of big networks and video editing suites – they were used as a way of storing large amounts of data in devices separate from, but connected to, the main PC. The price of external hard drives is falling all the time and at the moment you should be able to pick up a unit with 120GB for well under £100.

External hard drives are a great idea mainly because they can be used as back-up drives. You can keep all your programs on the main hard drive but dump all the important files, like emails or music files, onto a separate external drive, that can be switched off when you're not using it. These units connect to your hard drive in two ways – either via a USB connection or through the Firewire slot. Some better quality units (made by the likes of LaCie and Maxtor) feature both types of connection and if you have the money I'd go for one with 2 USB 2.0 slots and 1 Firewire.

Is it a complicated upgrade?

External hard drives are incredibly easy to hook up to your PC. Simply plug it in, power it up, and that's it.

Fitting an internal hard drive is a much more complicated exercise. If you do want to upgrade, you should have an open slot inside the computer to which you can connect a hard drive. It's a fiddly process installing a new hard drive and unless you've got some specific and very detailed instructions I wouldn't bother. If you have no open hard drive IDE connectors to accommodate a new drive, you can install an EIDE adaptor in an empty PCI slot.

> **note**: The absolute minimum for heavy downloading and ripping is 100GB.

DVD/CD

What are optical disk drives?

CD-ROM stands for Compact Disc-Read-Only Memory and, like a DVD, it's an optical disk capable of storing large amounts of data. The CD-ROM has replaced the floppy disk as the media for software distribution, as it's storage capacity means it could hold as much data as 700 floppy disks. In recent years, rewritable CD-ROMs (called CD burners) have emerged as the new standard. CD-R drives allow users to record information to a CD (compact disc), providing an easy way to archive data or share files. Additionally, you can choose CD-RW disks that allow you to write data to the CD multiple times. You might also encounter another variation on this technology, notably VCDs – these are video CDs that store video on them and can be played in some DVD machines.

DVD (Digital Versatile Disc) is the latest optical disk technology and can hold a minimum of 4.7GB of data, enough for a full-length movie. DVDs are commonly used to store films and other multimedia based materials. Similar to CDs, DVDs are rewritable but there are a number of different formats. In the main you'll encounter two rival formats: DVD-R and DVD+R (these allow you to record to DVD only once), plus the rewritable versions of each of these, DVD-RW or DVD+RW. CD-ROMs are still cheaper, but the price of both DVD burners and the media have collapsed in recent years.

Upgrading your optical technology – why bother?

Optical storage technologies allow you to dump multimedia content such as films and music onto a disc that can eventually be played on a stand-alone CD or DVD player. Also these optical storage technologies allow you to back-up to a 'physical' format – you can dump really important files onto something you can tangibly store away for future use.

DVD or CD?

CD-based technology is still cheap; CD burners cost less than £20 and the discs less than 10p. But DVD technology is fast closing in. Good DVD burners from the likes of Sony and LG cost as little as £35 and good quality DVD-Rs by the likes of Ritek cost less than 25p per disc. Rewriteable DVDs are still quite expensive with good quality brands like Memorex and TDK costing above £1 a disk, but these should last you for years. A standard CD can only store around 750MB while a DVD can store around 4.5GB.

But which format DVD should I go for and what speed?

Remember that there are a number of different DVD formats; ranging from DVD-R, and DVD+R, through to ROM and RAM. Don't worry about this confusing array of standards. More and more DVD burner manufacturers are producing multi-format drives that'll work with all formats.

The next issue is what speed your DVD machine will burn discs. CD burners can burn a disc in a few minutes (with 52x speeds) whereas DVD burners are still generally limited to 8x and 16x. Buy the fastest speed DVD burner you can afford.

I – Computer Hardware

> **Tip**: DVD burners will also burn CD media. This is particularly useful if you want to burn a CD of music that will play on a Hi-Fi. But remember that you have to have audio format files on the CD, not the main internet music format called MP3 (we'll encounter these formats in a later chapter). Also make sure that the speed stated on the packaging of your optical media – DVD or CD – can be burnt at the speed of your machine. It's no good having discs that say they can burn at 16x yet your DVD burner can only handle 4x.

Flash memory (solid-state removable storage)

What's a flash drive?

These are in fact what are called solid-state discs, which are high-performance plug-and-play storage devices that contain no moving parts. Most of you will probably have encountered one form of solid-state memory by now – flash memory, commonly found in digital cameras. Flash memory is small, light and fast. Due to its cost and capacity, however, it is used mainly in laptops, digital cameras, digital audio players, hand-held computers and video game consoles. Flash memory works a lot like your computer's memory and, like your computer's hard drive, allows you to store data almost instantly.

The formats?

Get ready for a long list. Every year seems to bring a new type of flash memory and most consumers are understandably confused about which one to use. The best thing you can do is to install some kind of multi-format flash card reader; that will have anything between 6 and 21 different little slots built into them which will read all the main formats. They can be installed either as a separate USB plug and play device (that means they don't need any special software) or you can buy PCI cards that can be fitted into your computer that have multi-card readers built into them.

The only real difference between all the different formats is size and cost. Compact flash is probably the biggest and cheapest while smartmedia and secure digital cards are smaller and more expensive. You'll also encounter some flash formats that are tied to particular manufacturers, like Sony – their UMD flash drives for Sony PSPs mobile games player and Memory Stick Plus are both very small and very, very expensive! The capacity of these various formats ranges from 128MB through to 4GB.

> **Tip**: At the absolute minimum make sure you have some way of reading Compact, Secure Digital (SD), SmartMedia and Multimedia (MMC). These are the most popular formats, although XD cards are gaining in popularity as a way of storing digital photos.

You'll also run into another format called USB flash drives. These are very small, portable flash memory cards that plug straight into a computer's USB port and functions as a portable hard drive with up to 4GB of storage capacity. USB flash drives are touted as being easy-to-use as they are small enough to be carried in a pocket and can plug into any computer with a USB drive, making them an excellent choice for file sharing and for use in small electronic devices. They're also automatically recognised by Windows XP Plug and Play system and so don't need any extra software.

Step 4 – Motherboard

What is it?

This is the heart of your PC. It handles system resources as well as core components like the CPU, and all system memory. It also accepts expansion devices such as sound and network cards, and modems.

Is it worth upgrading?

Unless your computer is more than five years old, the answer is probably not. If it is slightly older and a little slow, upgrading your motherboard could be worth it. A new motherboard will:

- allow you to use more or faster memory;
- support a faster graphics card;
- support the latest disc technologies like serial ATA or SATA hard drives

- enable you to make extensive use of peripherals that use USB2.0 and the latest Firewire standards; and
- increase your expansion slot potential.

> **Tip**: If your system uses a motherboard called Baby AT – dominant between 1993 and 1997 – you probably should upgrade to the newer ATX standard. This will make your computer work faster, and allow more USB and Firewire connections to be added.

What should I go for?

Most users like the bigger desktop and tower cases that are designed to use a standard ATX size motherboard, but more and more consumers are opting instead for small footprint computers (you'll also need a new case), and for these you'll require something called MidiATX or MicroATX designs.

What will it cost me?

Motherboard prices range from around £35 to over £400 depending on the features. Choosing the right board will dictate the type of processor you can use, how well the other components communicate, the features on offer and future upgrade potential. Cheaper boards can save you money but will limit processor support, graphics and expansion options. Spending around £70-£100 will get you a good mix of features and the latest technologies.

Is it a complicated upgrade?

It's a fairly complicated upgrade, which will probably end up with you replacing the case and all the other components as well. Unless you're really determined, I wouldn't bother – just buy a newer, cheap computer.

Step 5 – CPU

What is it?

The CPU (or central processing unit or processor) is the brains of your PC, as everything that happens inside your PC has to interact with it. It processes instructions, manages the flow of information through a computer system, and performs millions of calculations every second. The faster the processor (the higher the number of GHz), the more efficiently your computer will perform.

Is it worth upgrading?

If you're a big gamer or do a lot of demanding video and graphics work, you'll almost certainly benefit from the latest CPU – especially those with large amounts of L2 cache and fast FSB speeds.

What kind of upgrade should I go for?

There's a bewildering number of processors out there for all budgets, with prices from £25 for low-end CPUs up to £700 for the very latest and fastest speed beasts! If you regularly have lots of applications running at the same time, you will particularly see the benefit of Intel's Hyper-Threading technology. One of the biggest decisions centres on which brand to go for...

Intel vs AMD?

Intel's Pentium 4 processors are very much the market leader at the moment. But more and more people are opting for the 64-bit capability of chips made by a company called AMD, although the fact that it's the underdog in a global battle and its lower prices might have something to do with its growing popularity. Which particular type of product to go for though?

- While the cheaper Celeron or Duron ranges are absolutely fine for general tasks, if you are doing an awful lot of downloading, and starting to play a lot of music and video, it's basically a question of Intel's Pentium 4 versus AMD's Athlon XP and Athlon 64. The chief difference between these is that the Pentium 4 and Athlon XP are 32-bit CPUs while the Athlon 64 is a 64-bit model (we'll talk about all this 64-bit technology below).
- Dedicated gamers looking for speed and future performance improvements should choose one of the Athlon 64 chips. But be warned – they're currently very expensive! For heavy users with demanding tastes, Intel has also just launched a new ranges of processors, called Extreme Editions. AMD has reacted by introducing its own Athlon 64 FX range, a high-end gaming/workstation CPU; this amazingly powerful CPU can deal with one terabyte of RAM (1,000GB). Really demanding games require an awful lot of processing power and are designed to run more efficiently on a 64-bit platform.

> **Tip**: If you do decide to go for a very powerful CPU from Intel, go for the 'Prescott' models.

What's all this about the next generation of 64-bit processors?

It's all about the way data is transferred. The P4 generation (still dominant) of CPUs rely on something called the motherboard chipset to shuttle data between processor and memory, whereas the (Athlon) 64s avoid this bottleneck by integrating the memory controller onto the CPU itself. In simple language, that means these 64-bit chipsets allow you to work much faster with really data-heavy applications like graphics processing or video.

But you'll also need loads of memory (see below) and you'll also need to use software that has been specially designed to work using this technical architecture. Oh yes, and you'll also need a 64-bit OS to take advantage of this. As I write this, the technical jury is out on whether its really that great a technological leap but for the vast majority of users, it's not worth the bother at the moment.

Hyper-Threading – what on earth is this?

Most of Intel's current Pentium 4s and some of the Mobile Pentium 4s support something called Hyper-Threading (HT). Put simply, HT gives you two virtual processors for the price of one, which will give you a noticeable performance boost by allowing applications to use either or both of the two virtual CPUs, hence making the system much more responsive when running lots of applications at the same time. It's not quite as good as running a system with dual CPUs (as Apple does), but it's a lot cheaper.

Which clock speed to go for?

- If you use your computer for a limited amount of internet browsing, a small bit of music and plenty of word processing you could get away with using a CPU with a speed of as little as 500 MHz, although I'd recommend at least 800 MHz for better performance.
- If you're a more serious gamer who likes 2D games like strategy games, you'll definitely want a CPU speed of at least 1.6 GHz. The faster your CPU,

the better your PCs performance, especially for more advanced 3D gaming. This level is also for most average multimedia users and downloaders.

- If you're a hardcore gamer playing loads of top notch 3D games, or you edit loads of video or use your computer to make music, you'll want the fastest CPU you can reasonably afford, with a minimum of 2.4 GHz.

Is it a complicated upgrade?

On paper, the upgrade is easy. You simply open up the unit, whip out the old unit and put a new one in. In practice, it's a bit more complicated than that. You might have to upgrade the BIOS operating system – the software that lies beneath your Microsoft system. Unless you're willing to spend a bit of time preparing for the installation I wouldn't bother doing it yourself.

A good website that talks you through the whole process in enormous detail is http://users.erols.com/chare/cpu_gen.htm.

> **Tip**: Don't ever buy a CPU upgrade unless the new processor runs at least twice as fast as your current one. And if you're at all squeamish about opening your PCs case and pulling out parts, it's probably best to leave your chip alone. Also, increasing the amount of memory you have can also do as much to improve speeds as a CPU upgrade.

Step 6 – Graphics card

What is it?

A graphics card is a small circuit board (smaller than the main motherboard to which it's fitted) that contains the necessary video memory to bring your favourite programs, games and even video editing to life.

There are different types of graphic cards on the market – which one should you go for?

There's a huge range of models and types but they basically fall into one of three categories:

1. Gamer video cards

   These are the fastest and most expensive and pretty much essential if you're a keen video editor or into games like Half Life 2 or Doom 3. These usually cost at least a few hundred pounds.

2. Mass market video cards

   These are fine for most ordinary videos and games, are usually sold as something called AGP cards and they're at least 8 times as fast as a standard PCI card. They might also come as PCI Express (also known as PCI Express x16), which is the newest slot type and is as much as four times as fast as the fastest AGP slot. From 2006 this should pretty much be the standard for most new computers. It's also worth noting that a PCI Express video card will not fit into a PCI slot and vice-versa. Almost all PCs have PCI slots, whereas most PCI Express slots are found only on newer machines.

3. Value video cards

   These are for simple 2D games, basic video and music playback and more mundane business applications like spreadsheets or PowerPoint slides. These usually come as standard on cheaper PCs, but if you do need to buy one they should cost less than £100. These usually come as PCI cards – these slots are built into the motherboard and you can easily slot the PCI card in.

Any extra features to look out for?

- **S-video and/or TV-out**: Most video cards, even value cards, provide the ability to send the video signal from your PC to a TV. This is usually labelled as 'TV-out' and in most cases it's actually a connection called 'S-video' (you'll also see these on DVD and VHS players).

- **Dual monitor support** is also very useful for really powerful multimedia applications like video editing. Dual monitor support simply means that you split the video signal to go across two monitors instead of one. Most film and TV editors use this dual monitor system because it's easier on the eye, and allows them to run multiple applications at once. Usually the dual monitor support consists of one standard VGA output (standard computer monitor) and one DVI jack (see overleaf).

> **Tip**: If you intend to upgrade to an LCD monitor, you will need a graphics card with a DVI output port, as it will produce better image quality. DVI is a higher definition output, used with some newer monitors as well as some high-end TVs. If your monitor or TV supports DVI, using this jack will give you a better picture than the standard VGA or RCA (TV-out) output.

What's all this about the rivalry between ATI and nVidia?

The really important bit of a graphics card is the GPU or graphics processing unit. This is the clever bit that translates all the data into stunning visual images. And as with CPUs there are two dominant competitors: ATI and nVidia in this case. The truth is that both have made great products with nVidia (GeForce) traditionally viewed as the technical champ, but in recent years ATI's Radeon products have shot ahead.

Why bother upgrading?

If you are serious about using a lot more multimedia applications such as video editing, or imagine running more and more multimedia files over a home network, you should seriously consider upgrading your graphics card to at least the AGP standard. But beware: if you have a cheap motherboard it may have what's called an integrated graphics chip built-in instead of a proper, separate AGP graphics card. If your machine doesn't have an empty AGP slot but does have an open PCI slot, you may be able to add a PCI-based graphics card instead.

Is it a complicated upgrade?

The simple answer is: yes, in most circumstances. If your PC has an empty AGP slot, you can probably upgrade your graphics without difficulty; though if you have an integrated graphics board you may need to disable the system's on-board graphics chip in your PC setup program. PCI graphics cards are incredibly easy to fit.

Step 7 - TV tuner card or device

What is it?

It's a special kind of graphics card that allows your computer to receive TV signals, even digital TV signals, plus radio. It'll come with a slot to attach to your home aerial or its own small aerial. It will also (hopefully) have software that comes with it that will allow you to tune the TV signal, and even control recording of the TV programmes. These cards (or boxes) also frequently have inputs built into them that allow you to play MiniDV tapes and then convert them to data files. The price of these cards vary enormously, depending on the features (see below), but as a guide you'll probably end up spending around £50 for a cheap TV tuner through to £150 for an all singing all dancing Digital TV/DAB external tuner.

Any extra features to look out for?

The big feature to look out for is access to digital TV. If you plan to watch terrestrial broadcasts, a digital tuner compatible with Freeview will offer the best channel choice, picture and sound. Beware – in my experience reception can vary enormously and you may well need an extra indoor receiver specifically for the PC. Some tuners can also receive radio as well as TV broadcasts, but again beware – a digital tuner doesn't always mean you'll get Digital Audio Broadcasting (DAB) signals, which is a completely separate standard to simple digital radio. Also look out for Teletext and subtitle support.

Why bother upgrading?

1. There are a huge number of digital TV channels available with some great content – and they're all free. If you have a reasonably powerful computer with plenty of hard disk space you can, with the right software, record TV programmes and films and then store them on your computer for later playback. Digital tuners record signals straight to disk for maximum quality, while analogue signals need to be digitally encoded. The best analogue products offer MPEG-2 hardware for real-time video and audio encoding, but cheaper options rely on the PCs processor to do this in software.

2. Most analogue systems also allow you to capture video from a VCR or camcorder. Look for composite or higher-quality S-video connectors and stereo phono inputs. A good tuner can record material in Pal, NTSC and SECAM video formats. While quality can usually be adjusted, an hour of MPEG-2 video swallows up between 1GB and 2GB of disk space.

Is it a complicated upgrade?

Internal TV tuners require a free PCI slot, so you'll need to open your PC to install it but that shouldn't be too difficult. External options are much easier to set up, as they connect via a spare USB port, which is good for notebooks as well as desktop PCs.

> **Tip**: Although most are backwards compatible with USB1.1, a high-speed USB2.0 connection is often needed to use advanced features. Today's devices typically require Windows ME, 2000 or XP. Check specifically for Windows 98 or NT support.
>
> **Extra tip on the "EPGs"** that come with these cards: These are Electronic Programme Guides and they are software-based applications that tell your computer which programmes are due to broadcast. They also allow you to use your computer like a video recorder, setting up record times later in the day/week/month. But not all EPGs are created equal – some have primitive information and very few intuitive features. DigiGuide by contrast is top notch and worth paying less than £10 a year to access.

Step 8 – Operating system

What is it?

It's the software system that runs your computer and all the devices attached to it. In reality for most of us these OSs as they're called – operating systems – are made by Microsoft. This book is written for those of you with XP although Windows Millennium Edition (ME) is perfectly acceptable and should be able to handle most multimedia tasks talked about in this book. But ME is increasingly looking outdated and more and more software and new digital multimedia technology is specifically developed for XP.

If you have Windows 98, I'd upgrade fast. It's a perfectly acceptable system for bare bones systems that simply run a word processor along with the odd bit of email checking; but for data rich multimedia material and heavy downloading it's fast becoming useless. It's also worth saying that Microsoft has withdrawn live technical support for Windows 98.

You can also use a Linux OS but I wouldn't really bother unless you're willing to learn Linux code and feel fairly technically competent. It's cheap, it's resilient, it's open source and it's less vulnerable to security attacks by malware, but it's complicated and difficult to use. Apple make great computers and their Tiger OS is a wonderful piece of technology, but this book is explicitly written for Microsoft users – sorry!

Tip: If you haven't already done so, upgrade to XP now!

Is it a complicated upgrade?

Upgrading your OS from, say, Windows 98 to XP should be very easy. But there may be a few problems. First check that your computer has enough memory inside it to run XP – usually you'll need an absolute minimum of 128MB and practically 256MB. Also make sure that the programs you're running on Windows 98 can be upgraded to XP – you may need to update the software. Upgrading from Millennium is very easy.

Note: The absolute minimum for heavy downloading and ripping is Windows XP.

Step 9 – Sound card

What is it?

These days almost all PCs come with a sound card or sound chip – the technology that permits you to hear sound from your PC. Your sound card affects the quality of every sound your computer makes, whether it's a simple beeping noise or the music of a CD being played in your CD-ROM drive. Unfortunately there can be a noticeable difference in the quality of sound cards and many music fans choose to upgrade to a high-quality sound card, or to bypass their existing sound card using a USB audio adaptor.

Why bother upgrading?

A sound card should be fitted as standard, usually cheap ones! That's not necessarily a bad thing if you just use your PC for computing and downloading. If, on the other hand, you sit in front of your PC a lot and play music and watch film, you might need, something better. While some top range sound cards (they're built into the motherboard) will support six-channel (5.1) surround sound (two front, one centre, two rear speakers and a subwoofer), a dedicated sound card is always going to do a better job of bringing your favourite movies and games to life.

- If you're a **movie buff**, you'll need surround sound to play your DVDs and most add-on cards should fit the bill. Technical standards to look out for: a 24-bit card is probably the way to go for DVD movies sporting newer THX surround or Dolby Digital EX 6.1 or 7.1 sound formats.
- If you are into **creating music** then you may want to start with cards offering 24-bit 48KHz or 96KHz sampling quality with internal cards starting at around £60. The key make to look for in higher end cards is Creative and its Sound Blaster range.

Internal PCI cards are the cheapest option with some manufacturers offering respectable set ups for less than £20, although good makes like Creative will probably charge closer to £30. You can also buy external sound cards – these are easier to fit, typically sound better and have many more input and output slots. Entry level USB sound cards will cost around £35, but expect to pay £60-£80 for a decent 24-bit USB card.

> **Tip**: Look out for sound cards with something called S/PDIF in and out. S/PDIF stands for Sony/Philips Digital Interface, an audio transfer format that means you can connect speakers via a single optical or coaxial cable and enjoy much higher quality digital sound.

Is it a complicated upgrade?

Fitting an external sound card is incredibly easy – seconds rather than minutes. Fitting a new PCI-based card is a little more fiddly – you have to open up the machine as you would installing extra memory – but it's easy to do.

Step 10 – Home network

What is it?

Lets say you have three computers at home – one main desktop, a smaller older machine for the kids and a laptop. Unless you establish some kind of home network, these computers will not be able to talk to each other. And that means that the only way of transferring files between these computers is via some kind of disk swapping – floppy, flash or removable external hard drive. A home network of some form does away with this lack of connectivity plus you buy into a bunch of other big pluses!

1. The home network allows all the computers on the network to share files and potentially share printers.
2. It allows the computers to all access the internet through one machine – called Internet Connection Sharing or ICS – or through a special router.
3. You can also run other networked devices off the network such as streaming music and video players and even external hard drives that can be used by all computers on the network

How to build a network

There are two ways of building a network:

1. The first, and most traditional way, is by **wiring** them all using a technical standard called Ethernet which utilises 10/100 ethernet plugs and slots. It's fast, it's relatively secure and it works. But it involves lots of wiring and most home users tend to opt for the more expensive alternative.
2. **Wireless networks** use one of a number of technical standards to keep all the computers networked. These different formats – most of them grouped around something called Wi-Fi – all vary in the rate at which they can transfer data and their likely costs. As with virtually everything in techno-land, the bottom line is that the faster the system, the more expensive the solution.

Whatever technology you use, all the devices on a network will operate in one of two modes: **infrastructure** and **ad-hoc**.

In infrastructure mode, all network devices talk to each other through what's called a central 'Access Point'. This access point grants permission to each device, determines the frequency to communicate on, and relays data between network adaptors. Access points are stand-alone devices that can 'bridge' wireless computers to wired ethernet computers. They also may be built into other devices like network routers, which in turn manage the connection to the internet. Ad-hoc networks are easier to configure and only involve you in hooking up two Wi-Fi adapters, which then talk to each other on a peer-to-peer basis.

Which format?

- Up to fairly recently the main standard has been something inelegantly called **802.11b**. This format transfers data between computers at a data rate of 11Mbps (this is the absolute maximum) and its great advantage is that its become a de facto world standard.
- **802.11a** is a newer format and supports a maximum 54Mbps data rate and is fast becoming the de facto standard in some parts of the world. As a format it also has a number of advantages; it doesn't suffer from interference from Bluetooth devices like mobile phones and the data rate is easily fast enough to support streaming multimedia like film. Its big disadvantage is that it isn't approved for use in Europe!
- **802.11g** is fast emerging as the 'next thing' in Wi-Fi and also offers data transmission rates of up to 54Mbps, but unlike the 'a' standard it's backwards compatible with 802.11b – that means you won't have to junk all your old 'b' standard gear when you move up to 'g'. Its downside is that it's more expensive and only tends to work over shorter ranges.

What to go for?

If you're after a cheap wireless network with a wide range, go for 802.11b. But be aware that it can suffer from interference from other electronic devices (especially some cordless phones) and its data transfer rate will not handle any video streaming or extremely fast file transfer. If none of this bothers you the very cheapest option – at a cost of under £100 – is through two simple Wi-Fi cards or USB plug in devices. All you need do then is set up ICS or internet connection sharing, and you can then connect both of your computers to the internet.

If you want to connect up more than two computers you'll need an **access point** plus some adapters. At the moment it also makes sense to include a **router** in the black box if you can, as this can then handle all your internet connections. This raises the cost considerably – more than £200 for a network – and is also quite technically challenging, but is worth it if you want to build a robust and secure home network. Remember that routers also typically include some form of hardware firewall (see the next chapter on securing your computer) adding an extra level of security to your network.

> **Tip**: If you think there's going to be an awful lot of data traffic going over the network, especially if you have streaming devices that will handle video, you should go for a 802.11g network, preferably one with a router/access point. It's the most expensive option – it'll cost at least £100 – but it'll be technologically robust and should be able to handle all types of files including big film files.

Table 1.5: Computer connectivity options

Technology	Speed	Wireless	Range	Cost
Ethernet 10/100	100Mbs	N	A	A
802.11b	11Mbps	Y	B	B
802.11a	52/72Mbps	Y	C	C
802.11g	22/54Mbps	Y	C	NA
Firewire	400Mbps	N	D	A
Bluetooth	1.5Mbps	Y	D	C

Chart explanation:

- *Range*: In terms of a home network, will it cover the whole house? A 'D' means short distances, an 'A' means almost any home would be covered without additional equipment.
- *Cost*: A is the least expensive.

Is it a complicated upgrade?

Plugging in a couple of Wi-Fi adapters and installing the software for these adapters isn't difficult. Setting up the network and getting all the machines to talk to each other is considerably more challenging.

Windows XP, especially after the Service Pack 2 upgrade, does have some excellent wizards that help you set up a home network. These are a little complicated but worth the effort and shouldn't take more than 30 minutes to set up. Unfortunately you may also have to fiddle around with any firewalls you introduce onto the network – see the next chapter on securing your computer. The firewall has to be told to let your computer communicate with the other computer – this should be easy but frequently isn't.

Building a proper network based around a router is considerably more challenging and I wouldn't recommend this option except if you really mug up on the subject and are willing to spend a few hours fiddling around with network settings. Luckily, because most router access points involve a broadband connection (see below), they are usually installed by a friendly BT engineer who will do most of the fiddly stuff for you.

Step 11 – Streaming

What is it?

Streaming allows you to access music, videos and photos stored on your PC using a device which is in a different room – the device is plugged into your TV and Hi-Fi and has a remote control, just like a DVD player. It doesn't need to copy or store the files, as they're played or streamed over the network connection as and when you need them.

Why bother?

You could of course put a PC in your living room but for the vast majority of us there's neither the space nor the money to spend on this kind of media centre system. Streaming devices by contrast are a wonderful addition – I have a Netgear music player that sits next to the Hi-Fi and plays all of the 5,000 tracks that sit on my external hard drive. It's one of a number of sub £100 devices that lets you turn your PC into a jukebox that can be played in the living room.

The big downside with this technology is that it's quite fiddly to install – the operating software is fairly straightforward, but getting the device past a firewall can be extremely tricky. If you decide to go for a streaming device that also plays video, like the Pinnacle ShowCenter, you face an additional problem – many of the formats you've used to compress your films may not instantly play over the wireless network. Still, when these video streaming devices work, they are a miracle to behold – a video jukebox on your TV!

Step 12 – Modem and broadband

What is it?

A modem is a device that allows your computer to communicate with another computer over a phone line. You need a modem to access the internet. Most notebooks and desktops today come with an internal modem.

Is it a complicated upgrade?

Simple modems are easy to upgrade. If you want to upgrade, simply open up the computer box, unscrew the PCI card holding the modem in and install a new PCI modem card. It should take less than five minutes in all. You can also buy modems that connect to your computer through a USB port – these are trivially easy to install and should take less than a minute.

Which kind of modem?

Most traditional modems connect to the internet at a speed of 56K – this is the rate of data transfer between computers, and means that your connection allows up to 56 kilobytes of information to be transferred per second. Until just a few years ago this was seen as the standard, but now broadband speeds are the standard.

If you want to download lots of content off the internet, a broadband modem and connection is pretty much essential and DSL (or *Digital Subscriber Line*), is the most popular way of accessing the internet. It's faster, cheaper and to get it you don't need to install a special phone line.

You do, however need a special modem. DSL modems connect at any rate starting at 128Kbps right through to 40Mbps. It works by splitting your existing copper telephone line signal into two: one for voice and the other for data. But it only works if you're relatively close to the local telephone exchange – you usually have to be within four miles of an ADSL enabled exchange and the telephone local loop has to be capable of being upgraded. Initially ADSL speeds were set at about 512Kbps but recent technical upgrades have seen that mass market standard rise to a fast 2Mbps. Operators are also experimenting with services (using technology called VDSL) that provides speeds of between 4 and 16 Mbps.

What kind of equipment will I need for ADSL broadband?

The cheapest way to get online is using a **PCI DSL modem card**, with prices below £40. The downside is that it requires a fair bit of knowledge of how to install into your computer and also consumes a lot of your CPU cycle to operate, but it does deliver the best response times (gamers take note!).

A **USB ADSL Modem** is the easiest way to connect to a broadband network and should cost well under £100. It should simply be a matter of plugging in, installing the drivers and switching on.

More experienced users might want to consider an **ethernet/wireless router**, which is a stand alone device that maintains the connection to the internet for you. Most routers have an Ethernet connection to your local area network and act as a gateway, DNS, DHCP and firewall service.

Table 1.6: ADSL speeds

ADSL speed	Typical download	Typical upload
512Kbps	460 Kbps	220-260 Kbps
1Mbps	920 Kbps	220-260 Kbps
2Mbps	1840 Kbps	220-260 Kbps

Cable modems also allow users to access the internet via the same coaxial cable that brings cable television into their homes. This option is only available in some parts of the UK, but it has numerous advantages if it is available, notably:

1. It's usually **cheaper** than ADSL. With most DSL broadband packages you're looking at paying around £20 a month just for the broadband service, plus

extra for phone calls. Cable networks frequently provide broadband access, all phone calls and basic cable TV for between £30 and £50 a month.

2. It's **faster**. Cable operators tend to upgrade their networks quicker than BT (except that is for NTL in south London which has a lamentable technical record to date). So while BT has now upgraded its network to 2Mbps, operators such as Telewest are busily moving up to 4 and then 8 Mbps. There is one drawback though – if local traffic is heavy, speeds can fall back quite considerably, sometimes falling well below 1Mbps.

3. It's **easier to install**. A nice engineer comes and visits and installs a black box somewhere in your house, switches it all on, connects it to your TV and computer and then leaves you alone.

Speeds

Typical rate speeds vary depending on your type of internet connection. Dial-up users can get up to 54Kbps, but in practice typical connections are in the 40Kbps range. Cable modem users can expect anywhere from 1,000 to 3,000Kbps while DSL users typically get between 256Kbps to 2,000Kbps, depending on the access package purchased.

OK, regardless of which kind of network to use, what speed do I need?

You can never have enough broadband! If I were to switch into telecoms engineer talk, you could say that a 2Mbps connection will be *up to* 40 times faster than an old fashioned dial-up connection. But notice that *up to*. The speed your files transfer across the internet depends on many factors.

First off, some networks connect faster to the internet than others. But there's more. The internet itself is sometimes slower, sometimes faster – log on when the US logs on and internet speeds accessing websites will probably fall. And if your site is popular the speed at which you'll access it will also vary.

But back to the original question – how much speed do I need?

Most users will be perfectly happy with 2Mbps and I'd only really recommend going above this if you use downloading services like file sharing networks each and every day. If you do, you should be careful about any monthly download caps – these tend to be set at around the 2Gb level. A 2Gb cap will not last long if you download emails every day and maybe a couple of movies a month plus music. Pay extra and get that download cap removed!

Table 1.7: Internet speeds

User	Broadband speed	Likely cost per month (£)
Occasional internet user who only looks at a few pages a week, and checks their email intermittently	128 to 512Kbps	Under £15 a month although becoming increasingly difficult to find
Frequent emails, a lot of web pages and the odd photo moved across the internet	512Kbps to 1Mbps	Between £15 and £20 a month
Regular internet user but only limited downloads	1Mbps to 2Mbps	Between £15 and £25 a month
Very regular internet user, many with heavy attachments and some downloading	2Mbps to 4Mbps	Between £20 and £30 a month
Very heavy downloaders	2Mbps to 8Mbps	Between £25 and £35 a month
Streaming TV and radio over the internet – the cable TV experience but using the internet	2Mbps to 10Mbps (upper end for high definition TV)	Between £25 and £45 a month

Step 13 – Monitor

TFT versus CRT?

Good old fashioned cathode ray tube monitors (CRT) seem to have gone rather out of fashion in recent years, mainly because sexy looking flat screen monitors based on the TFT standard have crashed in price, but CRT technology is very robust and still has a lot of strengths. Yes, CRTs are bulky, heavy and consume a lot more power than TFTs, but they are cheap and they have an excellent reputation for reliability. It's also worth noting that CRT monitors usually have a higher resolution than a TFT of the same size, and work perfectly at any standard resolution up to the quoted maximum. Running a TFT below its native resolution (the number of pixels in the TFT) can result in either a very poor 'blocky' picture, or one that doesn't fill the whole screen.

CRT monitors

If you are going to go for a CRT monitor make sure it's one of the flat-screen models, which are much more pleasant to use than the old curved-screen ones. Most users will be happy with a 17in or 19in CRT, although larger sizes (21in and beyond) are popular for graphics-intensive work such as desktop publishing or photo editing.

> **Tip**: Ensure there's a smoothly operating tilt-and-swivel base fitted. A CRT should support at least an 85Hz refresh rate for any resolution you use to avoid flicker in the image. Cheap models might not achieve this at their maximum resolution. Check the data sheets closely. One of the most important standards for CRTs is the TCO label. This guarantees compliance with stringent electromagnetic emissions, ergonomics, energy saving and environmental requirements.

TFT monitors

TFT monitors are immensely popular and falling in price all the time. Decent 15in TFTs should cost less than £150 these days and some cheap 17in models are starting to fall below £120! In my experience anything below 17in is too small and the average is increasingly moving up to 19in as prices collapse. TFTs great advantage is that they're smaller, take up less room and look more stylish than CRT monitors.

You might also want to consider investing in a TFT monitor that can also double as a computer monitor. These have separate switchable TV tuners and PC inputs and they allow you to use the monitor for both purposes – they're still expensive, but if you're limited for space and you're looking for a really stylish looking unit, LCD TVs could be just right for you.

> **Tip**: Gamers wanting to run 1,600 x 1,200 will probably need an expensive 20in model – 19in panels are currently limited to 1,280 x 1,024, and budget 15in models might only support 800 x 600.

In this chapter we looked at the minimum hardware requirements for a computer to play music and video. We also looked at what steps to take if your equipment needs upgrading. In the next chapter we'll look at the steps you need to take to make sure that, when you go on the internet, your computer is secure.

Securing Your Computer

Be afraid, be very afraid...

No-one can hear you scream in cyberspace when your computer has been infected with a deadly virus. You can't cry on anyone's shoulder. And even if you have paid mega bucks for a fancy technical support line, they'll probably tell you to format your computer hard drive and start again from scratch.

And don't for one minute underestimate the challenge that's about to face your computer as you start downloading music and film – you are about to send your computer into potentially uncharted territory where viruses and nasty malware (a generic description for all sorts of bad stuff, malicious software) are endemic.

Let's get specific: going online and downloading content from the big file sharing networks is absurdly dangerous, unless you have *all* the following essentials:

1. a regularly updated anti-virus program;
2. a firewall, ideally at the hardware level plus one software based; and
3. at least two if not three strong anti-spyware programs.

If you've already got all these defences loaded up and ready for action, you're probably safe and secure – you can probably skip to the next chapter and start working with music. If not, read on!

In this chapter we'll walk you through a five step security plan that should save you future heartache and endless re-formatting. In this chapter you'll learn how to:

1. Make sure your operating system software is kept up-to-date.
2. Spot the enemy and understand the different types of malware.
3. Download anti-virus software.
4. Use not one but three great anti-spyware programs in combination as a triple dose spyware killer.
5. Work with hardware firewalls, ignore the Windows firewall and install the very best free firewall on the market, ZoneAlarm.

Five steps to safe downloading

Step 1 – Update your software

Operating systems, and especially Windows XP, have come a long way since the advent of 'always-on, always connected' broadband. XP has an in-built 'Update' routine that will automatically update your computer and protect you against any security weaknesses. These updates are not without their critics (who charge that some of the updates have made PCs even more unstable), but the protection these fixes provide from known 'nasties' (like the infamous Blaster virus that struck in the summer of 2003) is worth the risk.

Program profile – updating Windows	
User	Novice and experienced
What it is	Update Windows. It's a way of making sure your computer's operating system XP is up-to-date
Why bother	It'll defend you against any security breaches in your operating system
Source	www.microsoft.com
Difficulty	Very easy
How long will it take	A few minutes

Service Pack 2

> **note**: If you haven't already updated your Windows XP to Service Pack 2, do it now!

You can get the Service Pack 2 update as one very big download from the main Microsoft website, but in my experience it's a better idea to order one of their free update CDs. It'll take about two weeks to come out to you by post but it's worth having a hard copy of this huge update. The update is packed with loads of useful features but the best is the Security Center. This will install a much more proficient (but far from comprehensive) firewall, and it will make sure that any security software you're running is automatically detected and updated. It will also control the way that you continuously update the XP operating system.

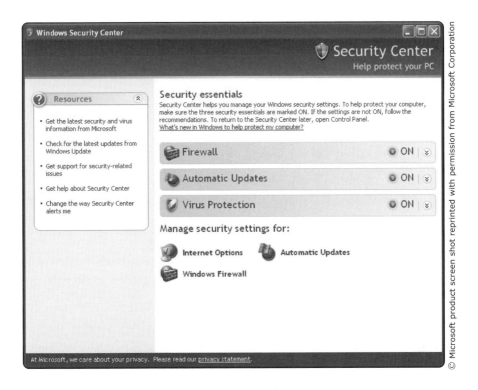

If you've not already done so, enable automatic updates in Windows XP with Service Pack 2. Right-click 'My Computer', click 'Properties', 'Automatic Updates', select Automatic (recommended), set a time for the updates (or accept the default), and click 'OK'.

Windows isn't the only place where automatic updates can save your bacon. Most antivirus programs also update automatically – all you have to do is provide an internet connection and keep automatic updates enabled in the program. Firewalls, too, occasionally suffer from flaws and exploits that require regular patching. ZoneAlarm firewall software usually notifies users when such an update is available. And at least one browser – the Mozilla Foundation's Firefox – notifies you of available updates. Other programs require that you check for updates manually via a menu command, or by checking the vendor's website for a patch or a new version.

Step 2 – Know the enemy

Even experienced internet users get downright lazy when it comes to security. They think they can just load themselves up with all the right programs, sit back and ignore the threat – the computer software will do everything for them! In my experience this is a hugely short-sighted strategy. I think it's worth getting to know the enemy via a judicious scan of press coverage to see what's out there and what you need to be careful about.

Malware

Malware (malicious software) – a term used to describe everything from viruses through to hacker attacks – changes every day and is constantly evolving.

A quick bit of history first. Malicious software first started appearing on dedicated networks such as the ARPANET way back in the 1970s. The boom in personal computers, initiated by Apple in the early 1980s, led to a corresponding boom in viruses. As more and more people gained hands-on access to computers, they were able to learn how the machines worked. And some individuals inevitably used their knowledge with malicious intent.

Malware comes in various different shapes and sizes. At the moment **network worms** are a particular problem as they spread very quickly over the internet or through local area networks. Their particular purpose is to penetrate remote machines on the network (yours), then launch copies of the program on these victim machines, which in turn spread further infections to new machines.

Classic viruses have gone a bit out of fashion in recent years, but they can still cause utter devastation on a victim's computer by launching and/or executing malicious code on your machine that then spreads throughout your computer wreaking havoc. The important thing to understand about these classic viruses is that, unlike worms, viruses do not use networks to penetrate other machines. Copies of viruses can penetrate other machines only if an infected object is opened (you initiate the program or code) and the code is launched by a user on an uninfected machine.

Trojan programs are also popular at the moment and are hugely varied and extremely dangerous. They perform actions without the user's knowledge or consent: collecting data and sending it to a cyber criminal, destroying or altering data with malicious intent, causing the computer to malfunction, or using a machine's capabilities for malicious or criminal purposes, such as sending spam. You'll hear a lot about the many different varieties of Trojans including Backdoors, Trojan clickers and even ArcBombs!

How do you know if you're infected?

Malware in all its gloriously horrible incarnations makes your computer behave strangely. Common symptoms reported to the experts at a great website called www.viruslist.com include:

- unexpected messages or images are suddenly displayed;
- unusual sounds or music played at random;
- your CD-ROM drive mysteriously opens and closes;
- programs suddenly start without warning on your computer;
- you receive notification from your firewall that some applications have attempted to connect to the internet, although you didn't initiate this;
- your friends mention that they have received messages from your address that you know you did not send; and
- your mailbox contains a lot of messages without a sender's email address or message header.

Viruslist.com also warns you to look out for 'secondary symptoms' including:

- your computer freezes frequently or encounters errors;
- your computer slows down when programs are started;
- the operating system is unable to load;
- files and folders have been deleted or their content has changed;
- your hard drive is accessed too often (the light on your main unit flashes rapidly); and
- Microsoft Internet Explorer freezes or functions erratically e.g. you cannot close the application window.

Not all these symptoms are necessarily caused by malware (badly installed software, faulty device drivers that need updating or severe hardware problems could always be the culprits), but always be on your guard when things start to go systematically wrong, or your computer starts to appreciably slow down – it could be malware! The simplest way of combating malware is to install a top range anti-virus program, but I'd also recommend a number of regular tasks including some or all of the following.

1. Every month check out the hit parade for top viruses

Most of the big virus software vendors run their own 'Top Virus' list (usually accessed via their website), but in my opinion Sophos (www.sophos.com) is the best. As I write this, variants of the Netsky worm are top of the virus hit parade. According to Sophos the Netsky virus accounts for just under 20% of all the virus' stopped by their software - in real terms that equates to hundreds of thousands of emails every day! That's an awful lot of nasty malware floating around out there. But it gets worse: according to another vendor called MessageLabs, virus infection rates are currently running at around 1 per 240 emails. By the time this book is published some new mutant nasty malware will have appeared and started wreaking havoc. Be alert and mug up on the enemy.

2. If there's a particularly virulent virus around, find out about it

Sometimes, just sometimes, you'll get a nasty surprise – your anti-virus security software won't stop a virus. It does happen, trust me. If this is the case you need to be smart and simply be aware of what's particularly virulent at the moment.

Let's take a current example – you might for instance want to mug up on a particularly horrid piece of malware I mentioned above called Klez. This is a mass-mailing worm which searches the Windows address book for email addresses and sends messages to all recipients that it finds. The worm uses its own SMTP engine to send the messages. It can also spoof the 'From' address in messages, a factor that has resulted in widespread confusion about the bug. The subject and attachment name of incoming emails is randomly chosen, making it harder for users to spot. The attachment will have one of the following extensions: .bat, .exe, .pif or .scr.

3. Take special care when you're downloading files from file sharing networks like BitTorrent

We'll encounter the big file sharing networks a little later in this book but one particular word of warning – malware designers are fashion conscious. If they know there are lots of people using one particular type of network or program they'll target them with specially designed nasties. Take The Nopir-B worm, for instance, which appears to have originated in France, and specifically targeted file sharing networks, pretending to be a program that allowed you to make copies of commercial DVDs. Of course it didn't and instead attempted to delete

MP3 music files on infected PCs. The worm normally arrived at the target computers as an executable file and activated when a user launched it.

Don't panic – what to do if your computer is infected

The helpful experts at Viruslist.com have compiled a handy 'get out of jail' checklist of things to do if you are infected and you don't have any anti-virus software installed:

1. Disconnect from the internet straight away.
2. Try rebooting using something called 'Safe Mode'. This is helpful if the CD/DVD player stops working. To access this: reboot and keep pressing F8.
3. Back up all important data to a CD/DVD, a floppy disc or an external hard drive.
4. If you can get online, try and download free anti-virus software and then run a full scan.
5. Alternatively you can go to Trend Micro's website and use its online virus scanner called HouseCall. See the section below.
6. You may also find that if you've been infected by one of the popular viruses or worms, the main anti-virus software vendors will also have a clean up tool available for free on their websites. The best way of finding out about this is to type the name of the likely virus or worm into Google and then add 'tool' and you'll usually find the specific clean up and detection tools available for free from the big vendors like Symantec and Norton.

Step 3 – Anti-virus software

You have a simple choice when it comes to anti-virus software – either pay for an all singing, all dancing anti-virus product from the likes of Symantec, or get a free program from an up and coming security vendor eager to build a reputation.

In most other markets the difference between the two would be huge, but not here – free anti-virus software is every bit as good as the paid-for variety. It may not have all the bells and whistles of the paid-for version but in my experience they work just as well – I've used the two free software packages AVG and Avast for over six years and I've only ever suffered one serious virus infection.

AVG versus Avast

Here's an oddity. Both of the free anti-virus software packages featured here, AVG and Avast, are Czech. Coincidence? I think not. If you're a small East European software company with lots of technical expertise but no global marketing the only way to build a global presence is to give your stuff away.

Program profile – AVG and Avast	
User	All
What it is	Free anti-virus software from either AVG or Avast
Why bother	They defend your computer against viruses
Source	www.avast.com or www.grisoft.com
Difficulty	Very easy
How long will it take to master	About ten minutes

The cynics amongst you may be wondering how they make any money. The answer is that although the software is free to you and I, commercial organisations have to pay a fee. The logic is that if your employees are familiar with the technology at home, they might just suggest using it at work. And it seems to work – go to big shareware and freeware sites like www.download.com and you'll see that millions of people have downloaded both programs, with 1 million new customers signing up to AVG every month!

Are they reliable?

The short answer is, yes. The main index used to measure an anti-virus systems' reliability is called VB100 and both have passed this key test.

The VB100 gold standard is published by a newsletter called the Virus Bulletin (owned by an anti-virus company called Sophos) and measures whether the program will detect all "In the Wild viruses" during a scan. Both Avast and AVG have passed the threshold for this VB100 standard in recent years (although they haven't passed the test every year in the past five years) and both are regarded as 'reliable' by the pros.

Which one to use?

Frankly, they're both very good. In my experience AVG:

- is easier to use;
- and has a nicer layout;
- but there have been grumblings from some users that its 'update' tool is not working too efficiently, sometimes leaving users exposed to old virus databases; and
- it can take a very long time to scan folders or individual files and is generally a bit slow; and
- you also don't get email support beyond Outlook built into the program.

By contrast, **Avast**:

- boasts a slightly more stylish yet complicated layout;
- does seem to be slightly better, and quicker, at spotting viruses;
- Avast surpasses AVG by scanning both Outlook and standard internet (POP3/SMTP) mail, making it a good choice if you use a non-Microsoft mail client; and
- amazingly, the company behind Avast, Alwil, also offers support via email for its free product.

> **TIP**: If you're new to broadband and slightly hesitant about computers use AVG, if you're more confident and worried about viruses get Avast.

AVG

Remember that AVG's developers Grisoft, like rival Avast, makes its money from satisfied users who graduate on to a registered paid-for version with full after sales support. That means they'll do their best to convince you to upgrade to the full or paid-for versions. In AVG's case that means that if you visit their main website at www.grisoft.com you'll struggle to find their free version – you'll be bombarded instead with chances to sign up for the paid for Pro version.

Installing AVG

1. Visit free.grisoft.com. There you'll see an option that says 'Get AVG Free'. Click on this and download the file.
2. You'll then have to install the program as you would any other (opt for standard installation).
3. You'll be asked to register your software by sending your email address to Grisoft. They'll then send you back an email with a registration key. Either take a note of the key or highlight the text and numbers, and then right click, 'Copy'. When asked to enter the key simply press 'Paste'.
4. Once installed, you'll see a quick First Run Wizard. This will ask you to update the file (which you should do and shouldn't take more than a minute), register for the free support website (you can do this later), run a system scan (which you can also do later) and create a rescue disk (which you should definitely do straightaway).
5. Once you've finished with the First Run Wizard you'll be sent straight to the main AVG front page.

AVG – Front page

How to configure AVG

Right click on the AVG icon in your system tray (lower right hand corner) and select Launch AVG Control Center.

AVG – Control Center

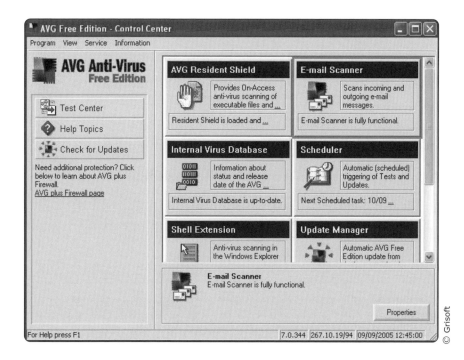

Select each of the tabs (Resident Shield, Email Scanner, Update Manager, Scheduler, Information). You'll then see a button with Properties – these control the settings on the program. Leave most of them at default.

It's worth checking the box that says Scheduler. This runs the automatic scans for viruses on your computer. Press on the button that Schedules Tasks and set the time to whenever you're unlikely to be using the computer. As standard the program will leave it at 8am. I prefer to run my tests at 2am.

If you leave your computer on most of or all day *and* have a broadband connection, use automatic settings.

When you do run a full system scan or choose to scan certain drives or files, you'll see the screen below appear:

AVG scanning

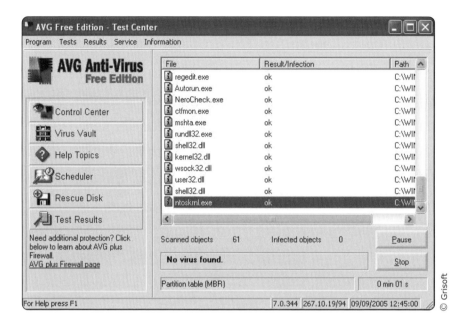

If it does find a virus you'll then see a box appear. This will clearly give you follow-on instructions for either moving the virus to a Virus Vault or deleting the malware file.

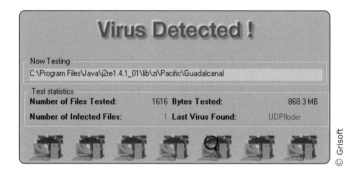

Warning – you're only licensed to use this software at home on one single computer. If you operate a home network it will not scan other computers on the network or even network hard drives.

Tips for using AVG

AVG, like Avast, works very well with the latest Service Pack 2 update for XP. But make sure that once you've installed the update the Security Center keeps telling you if the update facility of AVG – "Virus Protection" alert – is working. Go to 'Control Panel', select 'Security Center', and then select: 'Change the way Security Center notifies me' (select desired option).

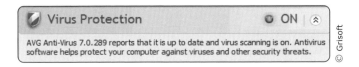

Due to the heavy server demand, updating AVG can sometimes be very slow. It's best to select an off-peak time for updates.

One particularly annoying feature of AVG is that it likes to 'Certify outgoing messages', which is actually just an advertisement for AVG and is not really needed. Go to the AVG Control Center, select 'Email Scanner', then select 'Properties', then 'Configure' and 'uncheck' 'Certify outgoing messages'.

If AVG detects a malware file in the System Restore folder it won't be able to remove it. Turn off System Restore, reboot and then run a full AVG system scan, and then turn System Restore back on and create a new 'Restore Point'.

Avast

You'll easily find Avast's free software on its main page at www.avast.com. Download the file – currently 2.5MB in size – and then install as normal.

When it's finally loaded, you'll be asked to register the program. You don't have to do this straightaway as you have sixty days to register through its website. Registration is free.

Configure the program for maximum protection – Avast's heuristics (scanning techniques used to trap new and unknown viruses) are set to medium sensitivity by default. To configure the program for maximum safety, click its system tray icon, then click on 'Start Avast AntiVirus' and then click on the settings symbol – the 'a' symbol – to open the on-access scanner's settings, and then move the scanner sensitivity slider to 'High'. The downside of setting an anti-virus program to its highest security level is that it might noticeably slow PC performance. If you're stuck with an older machine, you may need to go with the default security settings.

The lightning symbol controls the updates and the disc drive icon lets you start a systems scan or a scan of specific files.

You can control all the settings on the program by highlighting the symbol and then right clicking. This gives you the chance to alter the settings.

Avast settings

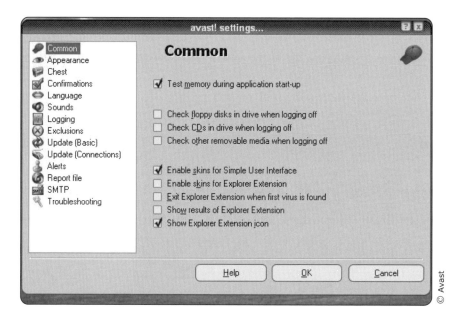

Avast has one great, hidden feature – it will run a system scan as a form of screensaver. This is a wonderful little tool and can be accessed by going to the control panel, selecting 'Display', then 'Screensaver'. You should see Avast as one of the options for the screensaver. Select and then everytime the system goes quiet Avast will run a system scan.

The alternative – an online scan

If you want to be absolutely sure that you are completely virus free, you can also run an occasional online scan with Trend Micro's HouseCall.

I always run this every month in addition to using Avast and especially after I've finished a particularly heavy round of downloading. It's free and easy to use – it installs a tiny bit of scanner software on your computer (which you have to download) that then runs a full system scan. It's not quick but it works – and it even offers you the facility to clear your computer of any malware infections. The website address is:

housecall.trendmicro.com

Program profile – HouseCall	
User	All
What it is	An online tool that scans your computer for viruses without installing a specific anti-virus program
Why bother	It's an extra line of defence for your computer to stop malware and should only be used in conjunction with a proper anti-virus program like AVG or Avast
Source	housecall.trendmicro.com
Difficulty	Easy
How long will it take to master	A few minutes

Trend Micro – HouseCall

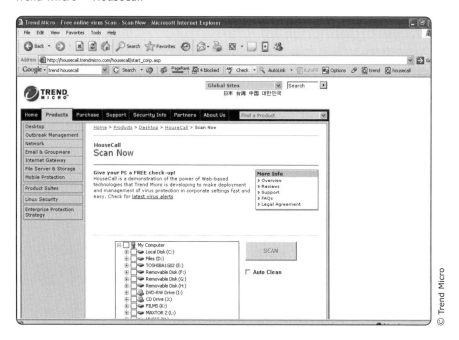

Once you've installed the small utility tool you'll be given the chance to select which drives you wish to scan. I'd recommend selecting all drives.

You should also make sure that the 'Auto Clean' option is ticked so that HouseCall cleans and deletes any malware.

To finish the test you obviously need to stay online throughout the scan.

Step 4 – Spyware

Spyware is annoying software that's installed without your informed consent and communicates personal, confidential information about you to a third party. The information might be reports on your web-surfing habits, or the software might be looking for even more sinister information, such as sniffing out your credit card numbers.

Sounds scary, but how does this software get on your computer? Look in the mirror – spyware is usually installed unknowingly by you, the user. 'Permission' is at the heart of the issue when it comes to spyware. Creators of spyware need your okay to install their products on your system. This is done by making any reference to the program as obscure as possible when you click on a site or install a program.

Here's how it might work. One particularly heinous spyware program called Gator, innocently offers to load a program that will help improve your internet browsing experience. If you click OK to the request, the Gator software will be installed on your computer and you'll suddenly notice an explosion of unwanted ads and targeted advertising.

Even more reasons why you should hate spyware

- It slows your computer down and it can even disable some programs from working properly.
- It hijacks your browser. This will cause your home page to be set to some site you never wanted (usually pornographic). It usually does this by adding a toolbar to Internet Explorer, usually full of spyware and viruses.

How to stop spyware

Task 1: Beware of ActiveX

The chief channel for transmitting spyware is through ActiveX controls conveyed over pop-up windows. A single click on a window that says "Click Here To Claim Your Surprise Gift" might give you a real surprise, by downloading spyware in the form of an ActiveX control (these are programs that are transferred across the net and then executed). Some pop-ups also contain a big button marked, "Click Here To Close Pop-up," or some similar text. Clicking on the button might actually activate a hidden ActiveX control that downloads spyware. Not all ActiveX controls are bad – Java based programs and websites make extensive use of the technology, but beware.

> **Tip**: Don't click on links within popups — pop-up windows are often spyware activators. Close the popup with the "X" on the titlebar and not the "close" link, if any, within the window. Also always choose 'no' when asked unexpected questions and generally be careful of unexpected dialog boxes asking whether you want to take a given action.

Task 2: Check out any unknown programs gobbling up system resources

Bring up the Windows Task Manager by hitting Ctrl-Alt-Delete and select the 'Processes' tab.

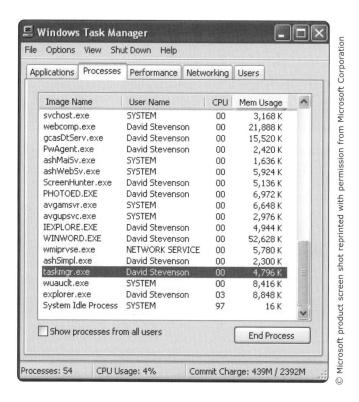

Click the CPU column twice and if you've got any spyware you're likely to see a few processes hogging as much as 90% of the CPU. The Image Name column may give you a few clues – beware anything with names like Gator, GAIN, QuickBrowser Update or TSA.

If you don't know what the image file represents, do an online search for the names that appear in the Image Name column using Google or your favourite search engine. If the processes are spies, simply click the End Process button for each of them.

Task 3: Install numerous anti-spyware programs

There are a lot of clever malware and spyware programmers out there who have a lot of time on their hands to design clever new programs that will spy on you and make them money. That means that spyware can prove quite difficult to remove, even for dedicated anti-spyware scanners.

If you're really determined to stop all the spyware on your system you need to install more than one program. Even the best-performing anti-spyware programs regularly miss dangerous malware.

Weapon 1: Ad-Aware

Program profile – Ad-Aware	
User	Beginner and experienced
What it is	A program that roots out spyware
Why bother	It stops people spying on your computer
Source	www.lavasoftusa.com
Difficulty	Easy
How long will it take to master	A few minutes

In my experience this is still the most powerful and easiest to use of all the anti-spyware programs available. Go to the company website at www.lavasoftusa.com/support/download/.

Install the program, making sure that the program has the latest update files loaded.

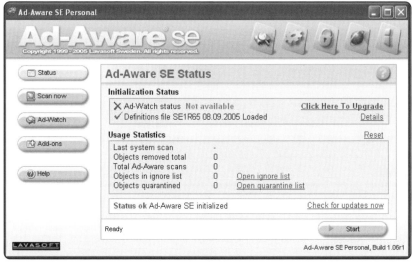

Hit the 'scan' button. A new screen appears – Preparing System Scan. Make sure that full system scan is selected for this first check and then press Next with all the default options.

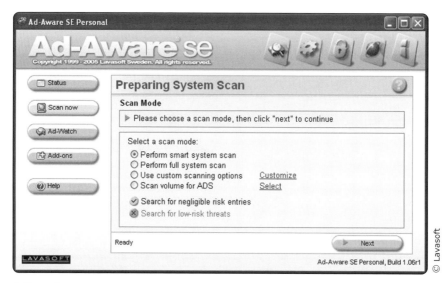

A Scan in progress:

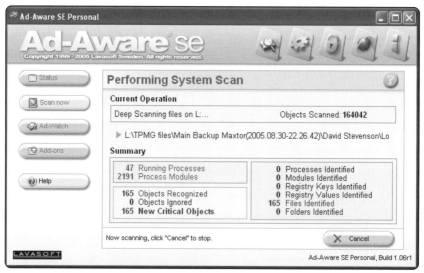

Once the scan is complete, you will be shown any suspicious files, registry entries or cookies detected. You can now delete or quarantine these files.

Beware about removing every single identified file. Some websites like eBay install perfectly harmless cookies that simply tell the site who you are; Ad-Aware will mix these up with malicious cookies that do need to be removed and destroyed. Carefully go down the list of objects and vendors and unclick anything you're happy with. Then press Next...

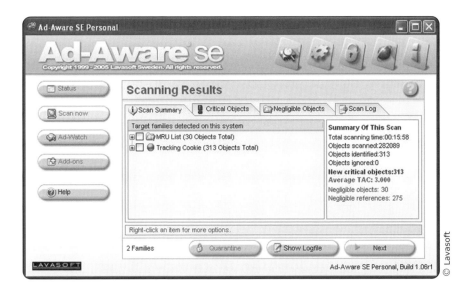

If Ad-Aware found any spyware, you should empty your recycling bin, restart your computer and scan again to make sure it is completely removed. Make sure to rescan your computer weekly.

Weapon 2: Spybot – Search and Destroy

Program profile – Spybot – Search and Destroy	
User	Beginner and experienced
What it is	A program that roots out spyware
Why bother	It stops people spying on your computer
Source	www.safer-networking.org/en/
Difficulty	Easy
How long will it take to master	A few minutes

In my experience Spybot works brilliantly in combination with Ad-Aware.

Install the program. Make sure you also check for updates while installing the program.

After installation your first step should be to immunise your computer – this runs a series of preventative measures that should protect you against future problems.

Spybot – Search & Destroy

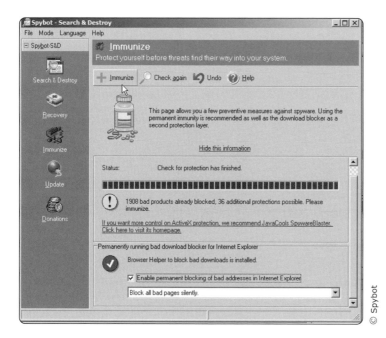

Your next priority is to run a full system scan. Click on the 'Search and Destroy' option and run a system checkup. Any problem software will immediately be identified and you'll be asked what you want to do about it.

Spybot – Search & Destroy scan page

That's it! Spybot is very simple to use. After you've immunised yourself and run a full system search it'll just keep running in the background providing you with constant protection.

> **Tip**: Find out what ActiveX controls have been installed. Service Pack 2 for Windows XP adds a nifty little feature to Internet Explorer 6 that lets you view detailed information about small programs called Browser Helper Objects (BHO) and ActiveX controls installed on your system. Many of these are enormously useful – they control how Java plug-ins are controlled or enable web-based software updates. But some of these BHOs and ActiveX controls are not so friendly and some may end up infecting you.

Luckily Spybot has a tool that lets you view detailed information about all installed ActiveX controls and BHOs and then get rid of the ones you don't want.

Go to Advanced mode. Click on mode in the top right corner of the screen and select 'Advanced Mode'. Then, click 'Tools' and then click on the 'ActiveX' and 'BHO' boxes, selecting both. You should now see them as options in the left hand column. Click first on ActiveX controls.

Spybot – Search & Destroy tools

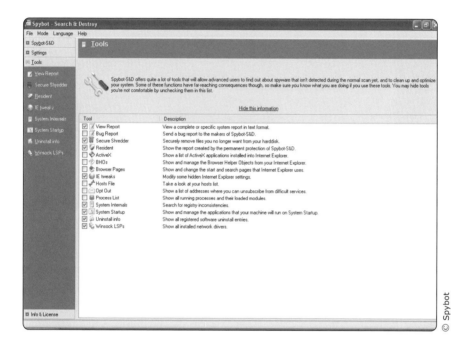

© Spybot

You'll now see a list of ActiveX controls installed on your computer. Those with green checkmarks next to them are known as legitimate and those with red signs next to them are known to be problematic. Those without either sign are unknown – if you want to check on them simply click on the line and you'll see further details listed. If you've still no idea what the program is, check it out at Google (type in a question) and if you're still worried click on the 'Remove' button at the top of this screen.

Spybot – ActiveX controls

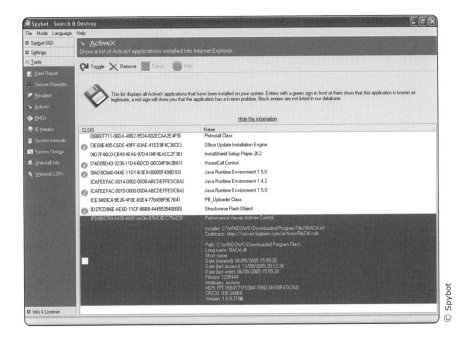

To view detailed information about a BHO or ActiveX control, just click it. If you think the object should be removed, click the 'Remove' button. Be careful – what might seem suspicious is in fact perfectly harmless or even essential to an important program! If you are not sure whether or not to remove an object, your best bet is to check through Google or ask a detailed question at an online help forum.

Another tool within Spybot's Advanced section lets you control which programs automatically come on when you start up your computer. In the Advanced Tool mode you'll see an option on the left hand column that says 'System Startup'. This shows you all the programs that load up when you start XP. If you have too many programs starting up it can sometimes slow the whole process of rebooting considerably.

Spybot – Search & Destroy System Startup

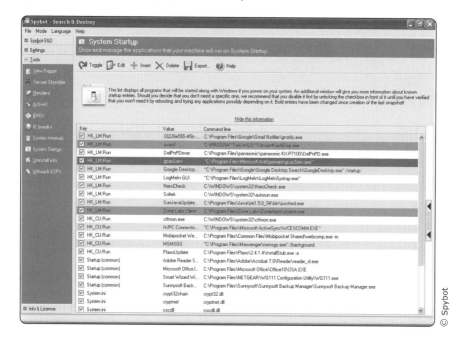

If you click on this Tool you'll see a long list of programs that are present at Start Up. Most of them will be fairly self-explanatory and even the ones you don't recognise are probably absolutely safe – if you want to find out what they are, simply copy the title of the file down and then Google it.

If you find something you don't like, simply click on the program (remembering to unclick all the others) and then press 'Delete'.

Weapon 3 – Microsoft's Giant killer

Program profile – Windows Defender (Beta)	
User	Beginner and experienced
What it is	A program that roots out spyware
Why bother	It stops people spying on your computer
Source	www.microsoft.com/athome/security/spyware/software/
Difficulty	Easy
How long will it take to master	A few minutes

A few years back a small software company started developing yet another free anti-spyware system called giant. It worked brilliantly, so brilliantly in fact that Bill Gates at Microsoft paid attention and started wondering out loud whether his Windows OS needed some kind of anti-spyware system like Giant. In true Victor Kiam style he decided that he did and decided to buy Giant rather than develop his own.

Microsoft's AntiSpyware – recently renamed Defender – is now available freely through the main Microsoft.com website and it's a wonderful piece of software. It integrates beautifully with XP and should be used in conjunction with both Ad-Aware and Spybot.

One potential, future, downside – it's still a beta test product. Microsoft haven't said yet whether they intend to keep providing the software for free and it's a fair bet that they might start charging for it (if they haven't already by the time you start reading this).

Beta – Front page

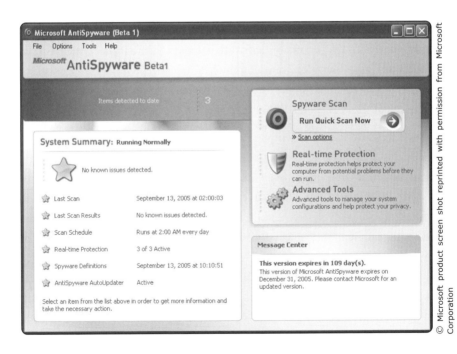

The main page itself is very elegantly and simply laid out. To run a scan simply press the 'Run Quick Scan Now' button.

Beta – Scanning

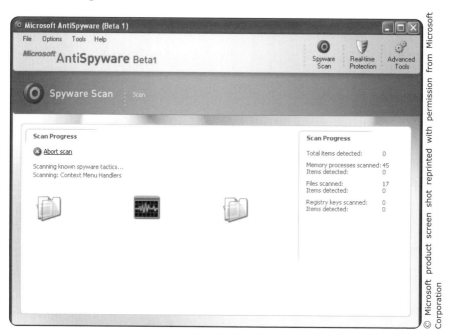

One very powerful feature of this software is its Real-time Protection function. This monitors all activity on your computer – and runs regular daily checks – and then cross-references any suspicious software with SpyNet, a global community of users.

Microsoft Defender also lets you view detailed information about any processes or programs that may be currently running on your machine – these could include spyware and other malware that are using system resources and generally spying on you. To use this feature click on 'Tools', and then 'Advanced Tools' and 'System Explorers'. Then click on the left hand column, underneath Applications, select Running Processes. You'll now see a list of process names and descriptions – click on a process to view more detailed information (if available). You can also stop a process from running if you think it's potentially dangerous. With the vast majority of processes you can click on them and in the right hand Application Details box you'll see 'This is a known process'.

Beta – Advanced Tools

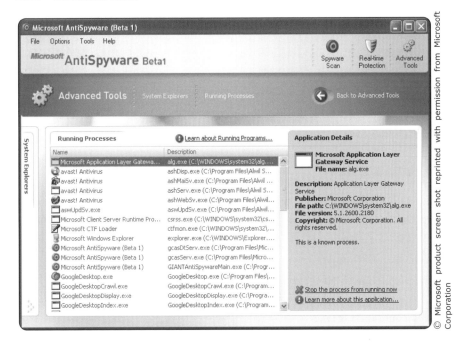

> **Tip:** This program has one extra, very useful feature. Click on the 'Advanced Tools' button and select 'System Tools' and 'Browser Restore'. This restores your internet browser to its normal settings if it's been hijacked by one of the many internet browser malware hijackers.

Step 5 – Firewall

> **Note:** A firewall is essential if you have a broadband connection. If you don't have one – **get one now**!

What's a firewall?

It's either built into your hardware (into the router for instance) or works through a software program, and limits access to your computer from a network and also limits your computer's access to other networks.

If used properly, a firewall gives you excellent protection against direct attacks from the internet, because computers' communication ports can't be seen over the internet if the firewall is set up properly. Many hackers use what are called 'open communication ports' to contact your computer and ultimately control it – if there are no ports available, naturally there is nothing to contact. That could mean your computer avoids being turned into a 'zombie machine' – this happens when a hacker gains subtle and non-intrusive control over a machine, ordering it on to the internet to perform certain specified functions.

What a firewall isn't

A firewall does not replace your anti-virus software or let you off the hook if you're entirely stupid and careless. It won't stop files on your computer being corrupted nor spreading of viruses and worms. It does, however, usually prevent harmful trojan horse programs and other backdoor programs from contacting the net and opening your computer to some hacker.

How do firewalls work?

Software-based firewalls such as ZoneAlarm are easy-to-use programs that allow you to configure the way in which communication requests to your computer and to the network are handled. Their strongest feature is that they allow you to easily set up a configuration that blocks all outgoing communications without your specific permission.

Your computer is vulnerable to attack in one of two ways: the file system and what's called the network stack, which is the 'protocol' that defines network communications. The file system, by contrast, is usually protected by anti-virus software that scans incoming email attachments and file downloads as well as inspecting files before loading, saving or executing them.

But your computer can also become infected whenever you're connected to the internet – even if your email program and browser are closed – through attacks against the network stack. Every computer connected to the internet has a unique address (the IP, or Internet Protocol, address) so communication can be directed to it. And with broadband, your IP address is always the same, so hackers can probe your computer at leisure.

Firewalls protect your computer's endpoints of communication (as opposed to the USB and other ports used to connect devices to the computer). Many internet services use specific ports – HTTP web traffic is usually on port 80 while file transfers (FTP) are usually on port 21. An open port can give hackers a way in, so firewalls close and hide all unused ports. Use of other ports is then governed by a clearly defined set of rules. A firewall, for example, may allow outgoing FTP requests but not incoming (so you can download files from the internet, but others can't pull files from your hard disk).

But what about the Windows XP firewall?

According to Microsoft, only 10% of its personal computer users have a firewall installed, a horrifyingly small percentage. Microsoft realised, bless them, that this represented a monumental security black hole and so resolved to do something about it. The answer? Their own firewall built as standard into their latest operating system. Windows XP now ships with its own perfectly respectable firewall. If you've upgraded your XP to Service Pack 2 this will automatically switch itself on, but if not you'll have to activate it yourself.

Go to Start > Control Panel > Network and internet connections > Network connections, then right click on your internet connection (which should be at the top of the page) and select 'Properties'.

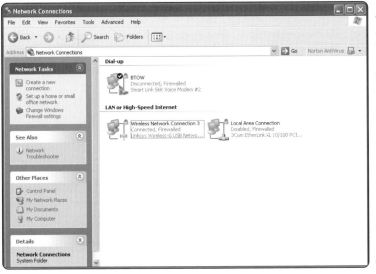

Now go to the 'Advanced' tab and place a check in the 'Internet Connection Firewall' box.

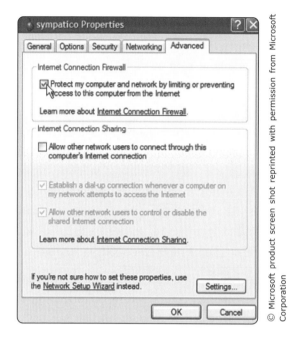

© Microsoft product screen shot reprinted with permission from Microsoft Corporation

Sounds great and free! Should I use it?

Microsoft's Defender anti-spyware software may be one of the best free security systems on the market, but its firewall is far from adequate. My advice is to get a better third party program, like ZoneAlarm. Why?

Windows' firewall has one very big, very important failing – it lacks outbound blocking, an absolute requirement for a good all purpose firewall. Inbound blocking – something which all firewalls (including Microsoft's) do – keeps illegitimate traffic from entering computers through your ports. But what inbound blocking doesn't do is stop a malicious payload from piggybacking on legitimate traffic such as email or web traffic going to Outlook or Internet Explorer. Once a malicious payload gets in, your inbound firewall kicks in. But malware designers are getting smarter by the day and not every nasty piece of code will be stopped by these programs – for that you need outbound blocking.

Also, the Windows firewall lacks one other crucial feature found on programs like ZoneAlarm – total lock down. This allows you to block all traffic if you think your computer has been infected or captured – any in or outbound network activity that isn't explicitly allowed by the pre-existing rules is blocked. Basically, there's no way to disable it unless you reboot the machine and uninstall the software.

Should I dump Windows firewall and use something else?

The short answer is, yes. Many of you may be tempted to run two firewall programs – the XP one and a third party one like ZoneAlarm – but my advice is simple, don't. Personal firewalls sometimes require a significant amount of tweaking and configuration to make them as secure as possible without impacting your ability to communicate on the network. Adding a second firewall on the same system can greatly confuse matters and make it exceptionally difficult to determine where any connection problems might lie or what you need to change to make it work. The simple answer is to switch the XP firewall off and install a proper dedicated third party firewall like ZoneAlarm.

And don't worry – XP will recognise the third party software and work with it. Software from the likes of ZoneAlarm is recognised by Security Center, and it will understand that you are in fact protected.

The firewall of choice – ZoneAlarm

ZoneAlarm is a hugely popular software-based firewall and justifiably so:

1. it's free;
2. it's easy to set up and use on a day to day basis; and
3. it works.

Program profile – ZoneAlarm	
User	Beginner and experienced
What it is	A software program that installs a firewall on your PC
Why bother	It stops hackers accessing your computer using the internet
Source	www.zonealarm.com
Difficulty	Moderately easy although running a home network can be very tricky
How long will it take to master	Ten to twenty minutes

Go to the main ZoneAlarm website at www.zonealarm.com and click on the 'Download & Buy' option on the top left hand side of the screen. You'll now see ZoneAlarms full range of programs – scan along and pick the free one called ZoneAlarm. Make sure you don't pick the paid for Pro version as it will disable itself after your one month free trial is up.

Install the program as usual. You'll now be taken through a number of different boxes, with registration boxes and configuration settings. One of the first asks for user information – your name and email. You don't have to bother with this as you can register later if you want.

You're then asked to fill out a perfectly innocent survey, asking you details of how you intend to use the firewall. Make sure you say you'll use the computer at home for personal use – any commercial usage is prohibited with this free version. After installation is complete, ZoneAlarm will probably try to encourage you one last time to take out the professional version, Pro, by comparing all the wonderful features missing on the free download. Hit 'Select ZoneAlarm free download' unless you do want to spend money on a proper professional firewall, which is by the way excellent and worth the money if you're a small business looking for property network security.

You'll now be given the chance to establish some settings for the program using the Configuration Wizard. First up, you'll be asked when you want to be alerted to any 'blocked' traffic. Personally I like to be informed at all times, so click the 'Alert Me' box. You'll also be asked whether you want ZoneAlarm to pre-configure access permissions – these are the rules you set that govern which programs can access your network ports. I like to set these manually so I know what's going on at all times, so select 'No, Alert Me Later'.

You're not done yet. There's still time for the Tutorial. It explains how the program works and is well worth the bother. A little under ten slides later you're ready to use the program.

Configuring ZoneAlarm

If you intend to become a prolific internet user and downloader it's really worth exploring how your firewall works, especially if you want to run a small home network sharing files between computers.

The first page is the Overview Page – this gives you a brief overview of the various programs that are allowed, the ones that are blocked and whether email monitoring is switched on.

ZoneAlarm – Overview

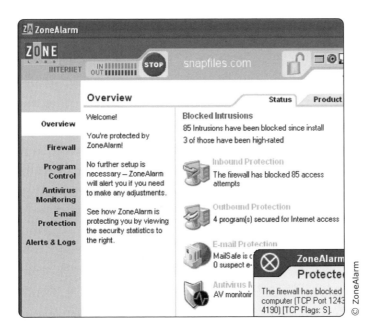

Click on the 'Preferences' tab at the top and check that ZoneAlarm has been set up to automatically update and that ZoneAlarm will load at Start Up.

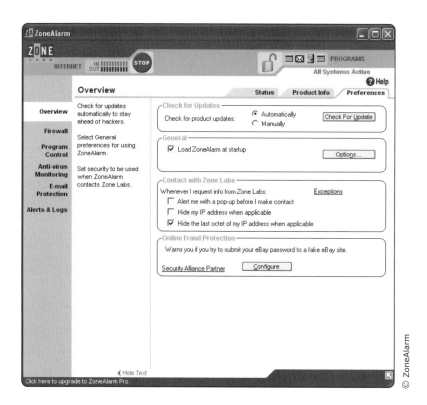

Next, look to the column of options on the left hand side and select 'Firewall'. Back at the top right hand side of the page you'll see two options: Main and Zones.

Let's start with Main – this is a hugely important page that allows you to dictate how secure you want to be on the internet. Your first options are for the internet and range from High to Low. Make sure this is set to High – this hides your computer's ports and disallows all sharing from your computer to the internet. Trusted Zone is for computers that you trust, like the ones in your own network, or if you have a shared network printer. Medium level is fine for Trusted Zone, but you might need to set it to 'Low' in some cases to able to use and share printers and files in your local network. The default settings here are quite safe.

ZoneAlarm – Firewall

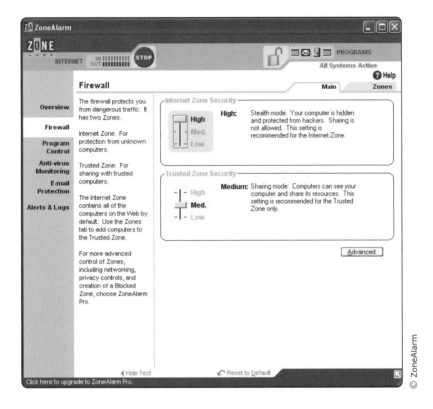

Now click on the 'Zones' tab at the top right hand side. You'll now see a list of the zones you've created that allow safe networking. This page allows you to configure your home network if you do decide to set one up. Unless you tell your firewall that computers are allowed to talk to each within a home network it will block all traffic. To enable a small home network simply click on the Add >>> button at the bottom left hand side – you'll be given the choice to choose from Host-IP site through to Subnet. To set up a home network choose 'IP Address'.

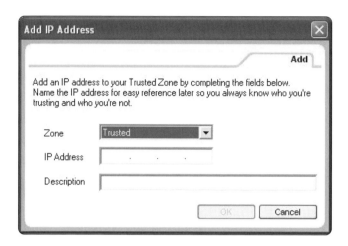

Select 'Trusted' – you're telling the firewall that the other home computers are 'trusted'. You now need to give ZoneAlarm the IP addresses of the other computers on the network. To find an IP address simply go to the computer you want to add to the network, click on its 'My Network Places', then go to 'Network Tasks box', and select 'View Network Connections'. If you have installed a wireless (or wired) network you'll see it listed (with a title like Wireless Network Connection 1 or 2). Double click on this icon and a new box called Network Connections Status window will appear. At the top of the box you'll see Support, click on this and then you'll see the IP address assigned to that computer. Now enter that address into the Trusted Zones box with a description and then hit OK.

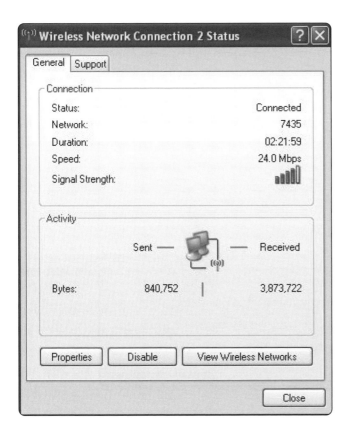

Your firewall has now been told to recognise the network computer for incoming and outgoing internet and file traffic. Going back to the Firewall/Zones screen, you should now see the computer listed as Trusted.

ZoneAlarm – Firewall program control

Now let's go to the Program Control box – look at the top left hand side of the screen and select programs. You'll see a list of all the programs that you have allowed to use the network or the internet. Whenever a program tries to access the internet – to update itself for instance – you should see a box appear asking your permission. If you recognise the program click 'Allow' and click 'Remember this setting', but if you are not *absolutely* sure this program is safe to be allowed to connect, you should click 'No'. You can change these settings later from within the Programs tab at the ZoneAlarm settings.

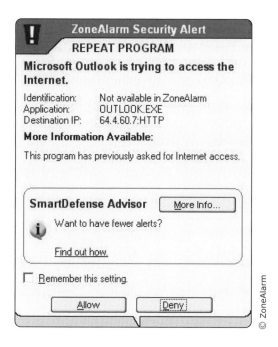

Once you've set all the permissions for different programs you should see the full list of programs in the Program Control box.

> **Tip**: Beware any program trying to get access to server rights. Never allow programs to have server rights, unless you really know what you are doing. Practically all backdoor programs (trojans) want to have server rights, so better to play it very, very safe here. It is likely that the program that wants server rights is a program that will allow other people to access the files on your computer. You can later change these settings from the Programs tab at the ZoneAlarm settings.

Double up your firewalls – use a hardware firewall as well

As broadband has taken off, so have sales of routers – devices for sharing an internet connection. These networking devices are installed between your cable/DSL modem connection and your home computer network and almost all include some form of firewall protection to filter the data that enters your network from the internet.

There are a huge number of different brands of routers out there, but most function in a similar way. The firewall can be configured by using a web browser like Internet Explorer to navigate to the router's built-in IP address (consult your documentation for this information) – the device contains a set of HTTP web pages which can be used to configure it.

This type of protection is known as a hardware firewall – it's a separate device that guards the entrance to a network, not an individual computer. Unlike software firewalls, such as ZoneAlarm, hardware firewalls are always active as long as the device itself is switched on. The only basic configuration necessary is to network your computers to the device correctly and enter your internet connection information.

The important point is to use a hardware-based firewall in addition to a software version. Because these hardware firewalls are external to your system, they can't monitor which programs are opening outbound connections, so they can't replace a software firewall running on the PC. However, they do keep incoming attacks off your local network, and they can shield your machine's IP address from the outside world, further protecting you from external attacks.

> **Ten point security action checklist**
>
> 1. Make sure Windows XP regularly updates itself. To do this make sure that Windows Updater is switched on.
> 2. Make sure you know every month what the top virus threat is on the internet.
> 3. Have a back-up plan in place in case your computer is heavily infected. Back-up all your key documents in My Documents to an external source every month.
> 4. Install all three anti-spyware programs.
> 5. Install either AVG or AVAST anti-spyware and make sure they update regularly and also run a full system scan at least every week.
> 6. Always be very careful of downloading ActiveX controls. See who provides the software or plug-in.
> 7. If you can afford to, install a router and a network. Hardware routers are an excellent security tool.
> 8. Install ZoneAlarm Free and make sure it's the very latest version.
> 9. Always scan any file you download from a file sharing network. Never open it up without scanning it.
> 10. Don't trust any attachments in an email. Always scan them before opening and if possible make sure your virus scanner has built in email support.

In this chapter we looked at the steps you need to take to make your computer secure. In detail, we looked at anti-virus software, spyware and firewalls. That's it! Your hardware should now be up to spec and secure. You're ready to start listening to music – the topic of the next chapter.

3

Digital Music

3 – Digital Music

The easiest form of digital, multi media content to master is music. Compared to video, music is easy to download, widely available over legal music download networks (we'll talk about this in our next chapter) and, in the great scheme of things, easy to edit and manipulate.

However, just before we go any further, let's linger on the meaning of the chapter title: *digital music*. In simple, layman's language, digital music involves taking music from the analogue world, usually in the form of a CD, and in some way ripping (converting) the music into a digital file. You'll hear a lot of nerds excitedly talk about this 'ripping', but don't be too confused: it's a simple term to describe the process by which the content on that CD is physically taken off the disc by a device like a computer CD player or some sort.

The resulting ripped content is then converted into a form of data that most computers can understand. As we'll see in this chapter, there are lots of formats these files can be converted into, but they all share in common some form of compression. This compression is done by complex mathematical algorithms and in the real world all you have to worry about is the quality of the conversion and the compression. Does the file sound hissy and noisy? Is the file of digital music too large to be put on a small capacity MP3 player?

The last step is to play the music. Which player to use is up to you – there's plenty of choice out there – but not all players are created equal. Some (such as the Windows Media Player) are immensely powerful as we'll discover, but a bit unwieldy and complicated. Others, like the jetAudio player, are less well known but incredibly simple to use. Pick the one that'll work for you.

In this chapter you'll learn how to:
1. Rip a CD onto your hard drive.
2. Play that music on your PC.
3. Convert that music into a smaller file that can be played on any PC or portable MP3 player.
4. Learn about different file formats including MP3, MP4, WMA, OGG and MPC.
5. Work out how to best use the popular Windows Media Player.
6. Record internet radio broadcasts.

In sum, this chapter is about three simple processes: ripping, compressing and playing.

Ripping software

Welcome to the world of Nero

Every once in a while we pesky old world types manage to beat the Yanks at their own game and produce a clever bit of world-beating technology. Nero is one such example – a CD and DVD ripping and conversion program produced by German firm Ahead.

Program profile – Nero	
User	Beginner
What it is	The very best CD and DVD ripper, copier and burner that money can buy
Why bother	It's super fast, and jam packed full of features and clever tools
Source	www.nero.com
Difficulty	Easy
How long will it take to master	Five minutes

Put simply, it's the king of CD/DVD burning and ripping software. It's software handles both the ripping and burning of discs perfectly and is worth every penny, although you can get it free with some DVD burners.

If it doesn't come with your computer, or DVD burner, you can try out the latest version (7 at the last count) for free at the main Nero site, www.nero.com. It gives you 30 days to give it a whirl before you have to buy. Download the file and then install it as usual (remembering to close any music player software you might have running at the time).

The easiest way to navigate around this very powerful program is to use SmartStart which will guide you through all the various options.

First tell Nero whether you're working on either a CD or a DVD. This is easily done – look on the top right hand side of the screen and you'll see either a DVD or CD. Click on the one you want.

Copying a music CD

Lets start with a nice and easy project – copying a music CD onto another blank CD.

Click on the star – Favourite – and then click on 'Copy Disc' (making sure the disc you want to copy is in the drive first!).

The sub-program within the suite called Nero Burning Rom will now come up. You'll see a main box giving you a number of options: these include Image, Copy Options, Read Options and Burn. You can use the preset options for most of the time – the only exception is if you want to keep an image copy of the CD on your hard drive. Why bother keeping an 'image' copy of the disc? Some people worry that they'll lose their CDs over the course of time (friends not giving them back, that sort of thing) and want to back up their favourite albums. The easiest way to do this is, of course, to copy the album onto another blank CD disc, but keeping an image file is a longer term solution. What this simply does is take a snapshot – an image file – of the CD and then save it to the hard drive. If this is what you want, click on the 'Image' tab and untick the little box that normally deletes the image file after the copying and burning.

When you've finished, press 'Copy' on the right hand side and away it goes copying the music. It may also offer you option of checking online with the big music databases like CDDB – these have a list of most of the albums and songs ever created and Nero cross-references the data on your disc with the online databases and then saves the ID3 tags to your computer.

The program now creates (very quickly) an image copy of the album. On my 24x speed CD/DVD burner it takes less than 3 minutes to create the whole image file.

When it's finished this part of the process, you're then asked to put in a blank disc. After you've done this, it then transfers the image file to the new disc, again in under 3 minutes.

When it's finished simply press 'Done' and the new CD pops out ready to use in any Hi-Fi system.

Ripping a music CD

Time to go back to the main StartSmart screen. This time select the main 'Audio' option (making sure you've still got the CD option selected) and click on 'Rip CD Tracks'.

Nero now takes you to the main Save Tracks screen. It'll almost instantly read all the tracks on the album and also present you with a small table that confirms the music data (album title, track names and so on) – select the right details and hit 'Selected CD'.

You can hit the 'Select All' button and all the tracks will be ripped, or just one track – you choose. You're also presented with the option of using different codecs (varying from Nero's own version of MP4) through the Windows Media files. And, finally, you can pick the location of the burnt file.

There's also a button on the bottom right hand side that says Options. This is mightily useful as it gives you a number of options you should definitely consider, including the option to remove silence between tracks, jitter correction and the option to automatically create a playlist of the tracks selected. I tend to tick all the boxes.

When you're ready just hit 'GO' and Nero will rip tracks in no time.

The Swiss army knife alternative – dBpowerAMP

Visit www.dbpoweramp.com and you'll see dBPowerAMP accurately described as the Swiss Army Knife of music ripping and burning. The most useful tool in an arsenal of handy programs is its Music Converter, a great way of ripping music on the fly and converting back and forth between different music codecs – it rips CDs incredibly quickly, is easy to use and it's free.

Program profile – dBpowerAMP	
User	Beginner
What it is	An enormously powerful program that will rip, burn and convert any form of digital music
Why bother	It's the quickest, best specified ripper on the market
Source	www.dbpoweramp.com
Difficulty	Easy
How long will it take to master	ten minutes

When you download the software on the 30 day trial you also get a PowerPack suite of extra features that includes a license to rip to MP3, various audio enhancement tools and an ID Tag editing service (though you do have to pay $14 for this pack after the trial ends). While at the website I'd also recommend downloading from the Codec Central sub site key audio codecs that are not included in the standard download – these include MP4 also known as AAC, OGG (aka OGG Vorbis), MPC or Musepack and the Windows Media plug-in.

You'll have to separately download these codecs and install them all one by one – it's all a bit time-consuming but well worth it. When you install the main program you'll see a number of initial installation boxes, and one in particular will confirm to you the various components installed and remind you that you're not registered yet. Just press OK!

At the end of this installation process (remembering to install the different codecs) you'll have two main programs: the DMC Audio CD program which does all the ripping and a separate dBPowerAMP Music Converter that lets you convert between MP4 and any other format.

The audio ripping software

Almost instantly the program reads whatever music CD is in your disc drive – like most of its peers it'll also go online to check the music details with the music database Gracenote (formally CDDB) or its open source rival FreeDB. At the top of the screen you'll then see the artist, the album and the genre.

You'll also see in the top left hand side a small screen symbol that says Rip. Next to it you'll see a small downwards arrow – click on that arrow and you'll see a small menu come down. Select 'Rip with Options'.

dBpowerAMP – Rip screen

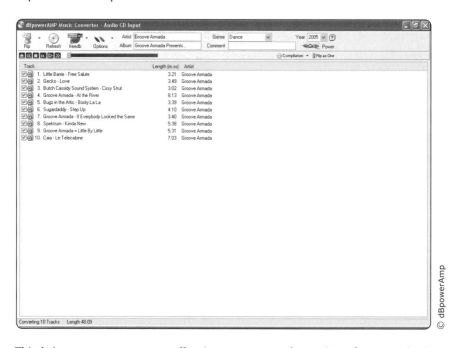

This brings up a new screen allowing you to set the options for your ripping. The most important option is to work out which kind of music codec to use – most of us will probably end up using MP3 or WMA, but there are some very powerful alternatives lurking around (see the next section on which codec to use). There's also a slide bar that lets you determine the quality at which the music data is imported. Typically with MP3 the very best results are above 300Kbps, the worst below 100Kbps. If you're worried about the size of the music file and want to play it on a portable music player with only small storage capacity (below say 500MB) then select 128Kbps or even lower.

You're also given the option of changing the frequency rate of the sampling – make sure it's set at 44100. You can also change the location of the saved file in this box. When you're ready, hit 'Convert' – a few minutes later the music will have been ripped from your CD onto your hard drive.

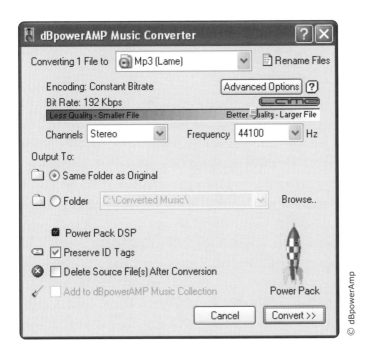

Codecs galore – which ones to use?

The term *codec* comes from a combination of the words compression and decompression. The name of the game with all codecs is to squeeze down, or compress, digital music into small file sizes without losing too much sound quality.

All the clever compressing and sampling is done by using horribly complex algorithms that profile the music track and take out bits of sound that aren't necessary. How each of the codecs does this varies enormously. Some codecs, such as FLAC, are described as lossless because they retain the true multitude of sounds in a given audio track, but these lossless codecs tend to produce much bigger files. Lossy codecs, by contrast, pick up all the main elements of the track but don't sample every little nuance, and the size of the resulting file is much, much smaller.

Luckily, most ripping software packages – dBpowerAMP, Nero or Media Player – will give you the chance to save ripped music with a choice of codecs, although most people will use either MP3 or the Windows WMA file format. But there are some other formats worth exploring.

MP3

In techno speak, this is known as 'MPEG-1 Audio Layer-3' and it's by far the most popular compressed audio file format. An MP3 file is about one tenth the size of the original audio file, but the sound is nearly CD-quality. The system works by analysing the waveforms in the sound file and ignoring those that the human brain is poor at perceiving, and then compresses the rest. Because of their small size and good fidelity MP3 files have become a popular way to store music files on both computers and portable devices. It's worth noting that many digital aficionados prefer a version of MP3 called LAME, which is an open source version of MP3 (you can download the file at lame.sourceforge.net). The consensus (in 2006) seems to be that the LAME version produces the highest quality MP3 files at average bitrates of 128 Kbit/s and higher.

WMA

WMA files are Microsoft's take on audio compression with a few added twists. Its big advantage is that it can compress music to even smaller files than MP3 without a significant decrease in sound quality. It also integrates easily with most major music players like Media Player and jetAudio (we'll talk about these later in the chapter). The downside of WMA files is that they don't always play on portable music players and it's also heavily used by music websites that like to build in some form of copyright protection into their music. We'll talk about this copyright malarkey plus DRM – Digital Rights Management – in much greater detail in the next chapter, but suffice to say that protected WMA files can stop playing if the copyright holder decides you've violated their license terms (i.e. by playing it on a device they're not happy with).

MP4

MP4 or M4A or AAC. A worthy successor to MP3, MP4 is a container that can store many sub-formats, of which the most popular (certainly amongst Mac users and iTunes fans) is Advanced Audio Compression (AAC). You might also encounter another format called .M4A – Apple's lossless format (ALAC) – which is a bit like a zip file in that you lose no sound quality in the compression. There are also .M4B files, which are identical to .m4a except that iPod knows to use them as audio books and allows book marking.

MPC

MPC or Musepack/MPEGplus is the king of lossy encoding. 'Lossy' as a term applies to any compression technique that 'loses' some of the audio file in the compression. Musepack works brilliantly at compressing but takes next to nothing out of the audio track – the only downside is that the files can be quite big and most portable music players like iPods won't play the format.

OGG

OGG, or OGG Vorbis, is a very popular free encoder, which produces compressed files that are of an appreciably better quality than MP3 and WMA, equal to M4A (AAC) and nearly, but not quite, as good as Musepack files. The resulting files tend to be larger than MP3 files, and smaller than Musepack files plus OGG has one last advantage – a growing number of portable music players will recognise the format.

FLAC

You might also run into yet another format called FLAC. This is a lossless compressor – this means that nothing is thrown away, unlike MP3, WMA and other similar lossy compression methods. FLAC files will only compress to a maximum ratio of 4:1 and a 3 minute track will take up about 30MB.

How do they compare?

If you were about to pump your prized digital music through a ten thousand pound Hi-Fi system you might care enormously about the sound quality. In which case, the FLAC and Musepack formats would end up being the undisputed winners. But these high quality codecs also produce large files and in most cases are, frankly, overkill.

I, like most people, use compressed music on my home computer – with rubbish speakers – and on the move with cheap portable music players and knackered headphones. The quality of the music is important but nowhere near as important as how many tracks I can fit on the music player. In these circumstances the heavily lossy formats like MP3, MP4 and WMA will tend to be quite satisfactory.

> **Tip**: Make sure you set the sample rate (in Kbps) at a satisfactory level. For most people that should be above 64Kbps and for best results above 128Kbps.

The codec you use also depends on your PCs music player and your portable music device. Some only accept MP3 formats, although a few smaller manufacturers also sell machines that will play OGG files. But let's assume for arguments sake that you do manage to find a music player that accepts all these lossy formats – which codec to use?

The simple way of testing the efficiency of different codecs is to compare their compression rates. Let's take a typical music track, in this case the superlative *Grand Love Story* by Kid Loco, and see how much we can compress the music track without appreciably affecting the sound quality. For the record, the original length of the track is just over 5 minutes and on the CD the audio track takes up 41.866MB (uncompressed).

Codec	Sample Rate / Frequency	File size in MB
WAV (original)	Uncompressed	41.866
WMA 9.1	192 kbps / 44 Hz	5.73
MP3 (LAME)	192kbps / 44 Hz	5.69
MP3 (LAME)	64kbps / 44 Hz	1.9
OGG Vorbis	192 Hz	5.86
M4A / MP4		3.316
Musepack		7.604

Verdict

The MP3, 64 kbps version sounds a little ropey and hissy, although the file at 1.9MB is incredibly small, while the MP4 track sounds great and is only just under twice the size. The best sound of all comes from the Musepack file with OGG trailing a close second. The best all-rounder is probably the WMA file, which is relatively small (at 5.73MB) and still sounds great. But be aware that different styles of music might produce different results. Some codecs, like MP4, sound appreciably better with lighter classical music tracks for instance – and so the differences can be marginal at best. The bottom line is that MP4 produces good quality, very small files, while OGG produces a lovely rich sound but much bigger sized files – the best all-rounder in terms of both quality and file size is probably the WMA format, which produces good sound quality in relatively small file sizes.

Digital music players

Windows Media Player

Now that we've learnt how to rip music from a CD, it's time to start playing our digital music.

The default media player (it also plays film and music) has to be Windows Media Player, currently in its 11th incarnation. If you've got a PC with Microsoft Windows on board – XP or Millennium Edition – you'll almost certainly have it installed, even if you don't actually want it!

Bundling a media player free with Windows, however, is not necessarily altruistic on the part of Microsoft. It's all part of a very clearly defined strategy by Bill Gates and his bespectacled minions to lord it over the world of digital content by providing the default media player of most PCs. If all that sounds a bit paranoid then consider the various huge legal actions by competition authorities in both the US and Europe to stop Microsoft 'bundling' up Media Player in its suite of software.

There are more than a few sceptics who wonder whether Media Player is little more than just a handy media player with a bunch of features that make it useful, but far from indispensable. As you're about to discover it's great that you get these features for free, but there are better alternatives available for the more discerning digital music fan.

Still, despite these reservations, it is free and very easy to use, and almost certainly sitting on your computer ready for action.

Program profile – Windows Media Player	
User	Beginner
What it is	A sleek, powerful media player that works with film and music
Why bother	It's the biggest media player freely available on all PCs with XP
Source	www.microsoft.com
Difficulty	Very easy
How long will it take to master	Ten minutes to half an hour for all the features

Tip: Get the upgrade to the latest Media Player right away. Formats, codecs and content are changing at an astonishing rate and most media players struggle to keep up with all these changes. The only fix for this problem is to upgrade to the latest version as quick as possible. And this is certainly true for Media Player – get the latest version as soon as possible by visiting its sub-site at Microsoft at www.microsoft.com/windows/windowsmedia/. Here you'll encounter one pleasant and one unpleasant surprise.

The pleasant surprise

To download the latest version you don't have to run a small bit of software that takes an inventory of your existing software and then forces you to download loads of confusing options – as is the Microsoft way. Instead, you:

- Click on 'Download'.
- Save file to a place on your disk.

The unpleasant surprise

It's a huge file. The last time I looked it was 12.1MB in total. Even on a fast 1Mbps broadband connection that's a long download – probably ten minutes or more.

What to do with the downloaded file

First off, make sure you're online when you install the download – the software will need to go to the internet to complete various registration checks and activate some add-ons.

Double click to launch the installation. Hit 'Run' when the Open File – Security Warning comes up.

The software will now extract all the relevant files and then present you with the main install window.

Windows Media Player – Welcome screen

Click 'I Accept'. The installation program immediately tries to go online to check the version of the software. After this it starts installing all the files and then it's time to set your preferences.

3 – Digital Music

Windows Media Player – Welcome screen during install

Click 'Next'.

You're now presented with a number of very important options. Most of them are straightforward. Tick 'Yes' for the option to display information from the internet (it's a nice feature that checks the data about each media file), and tick 'Yes' to send the file and URL history to the player.

Tick 'No' to Acquire Licenses automatically for protected content. You don't need to have this checked and, frankly, it's none of Microsoft's business what you play on your machine.

Tick 'No' to send Unique Player Id to content providers. Why should you voluntarily sign up for someone snooping on your music collection – I refused and so should you.

Selecting a default media player

The next screen is also hugely important. It lets you set the default preferences for various media formats – music and video. You can, at Microsoft's prompting, tick them all. This means virtually all major media files will automatically load up Media Player. If you intend to carry on using Media Player exclusively, for eternity, click them all. If you'd rather use more specialist players for different types of content, unclick them all. Remember, whenever you load digital content your operating system will ask you which program to use anyway, so you'll always have a choice. Personally my preference is to click 'Yes' to any files that are based on Windows Media formats (like WMA or WMV) plus the AVI and more obscure files like Midi or AIFF. But in my experience there are better programs that play DVDs and CDs. The choice, though, is yours.

Click 'Finish'.

An icon will probably now appear on your desktop and Media Player will go online again to validate your settings (if it does and your firewall comes up asking you for instructions, let it through and click the box to remember this instruction).

Time to start playing with Media Player

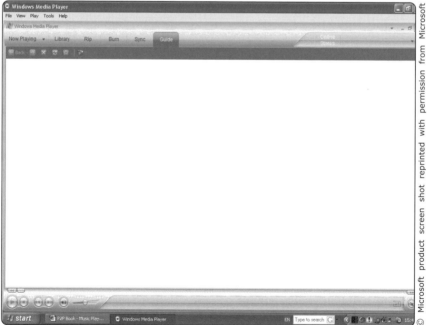

The Main Media Player screen

The Library – setting it up

Most newcomers to Media Player are advised to start off with the Library.

Look on the top left hand of the screen and you'll see Library as a tab.

When you first go to this section you'll be prompted to search your computer for files. You don't have to do this (you can just launch this file searching process through the tools menu) but in my opinion it's a useful feature, so say 'Yes'.

If you have drives both inside your computer and externally (Firewire or network drives for instance) you should probably specify All Drives on the next screen, which asks you where to search. Then click to start the search. This clever little utility now goes away and searches through all your drives and indexes all your music files, building up a formidable database of content.

Using the library

Your next priority is to organise your music collection so that you know what's where. The good old-fashioned way was to simply put all the music files in the right folders and then order them alphabetically in a big section called MUSIC somewhere on your hard drive. Media Player's Library offers a typically elaborate alternative to this – it builds a huge database of files, which are then organised via their ID Tags. There are four main ways of categorising digital music.

- Title
- Artist
- Album
- Genre

The first three, I think, are fairly straightforward but the last is a curious parvenu. Most American software and websites love this curious term, genre – it's supposed to help guide the poor consumer through the tricky business of what's Rock, or more specifically Alt Rock, or even more specifically Alt adult Oriented Rock with an Armenian trip hop flavour. Frankly, I find the whole thing a bit silly as most of the music I like doesn't fit into any nice, easy to define category (somewhere between LoFi, Trip Hop and Progressive Rock apparently), but still it's obviously useful as millions of people seem to organise their music collections on this basis. I organize my music by clicking on the heading which says Album – this automatically sorts all the content by the album name.

Playlists

The option of using the Library function to scour your music collection and then organise it also brings with it one other advantage – it organises your playlists.

Playlists are handy little files that tell a media player to play a certain number of music files in a certain order. Most good ripping software will take the music on a CD and then dump down a little playlist text file (many are called M3Us) which organises the music. That means when you play the music on, say, a portable MP3 player, the playlist file will tell the player to run the files in the correct order.

Playlists are also useful if you want to generate new compilations and Media Player has a handy playlist function built into it. On the left hand side of the main player window you'll see a series of choices starting with 'All Music'. Further down you'll see 'My Playlists'.

- If you want to build a new playlist compilation Right Click on your mouse and you'll see an option that says New. Click on New.
- On the far right side you'll see a new box column appear headed New Playlist. Inside this box you can now place your selected music files in whatever order you want. The simplest way to do this is to look back to the left hand side again, click on all music, pick out tracks for the compilation and then drag and drop them inside our New Playlist box.
- When you've got all your files in the right order you can then play them all (holding down 'Shift', click on all the files and then right click 'Play').
- Alternatively you can save this playlist for future use, say at a party. Put the cursor over the box at the top that says New Playlist and then click. You'll see a number of options one of which is Save Playlist As. This allows you to save the playlist in whatever format you want (Windows Media Playlist or M3U). The default storage location is probably the best though – inside a file called My Playlists inside My Music which is in My Documents.
- Want a two hour long party playlist? All you have to do now is build the compilation, save the playlist inside a selected folder and then right click the playlist file itself and hit 'Play'.

3 - Digital Music

The playlist alternative - DioneSS Playlist Editor

Playlists are an absolutely indispensable tool for serious music lovers. They save you the bother of having to constantly click play for each music file you want to listen to and they also let you build up some very cool compilations. Every major music player like Microsoft's Media Player and Apple's iTunes (we'll run into this in the next chapter) will use and compile a playlist of some sorts. But they're not necessarily the most efficient way of batch compiling dozens of playlists in one go.

Thankfully, there's a great, free alternative that can quickly, and easily, build playlists. It's called DioneSS Playlist and it's available at http://homepage.ntlworld.com/thorin92/dss/. As freeware goes it's wonderful, mainly because it's not made by Microsoft or Apple, doesn't care what files or formats you want to use (even if they're called weird things like OGG) and is incredibly simple to use.

Playing music

Now that you've organised your growing collection of digital music, it's time to actually play the stuff.

Media Player is admirably straightforward - the easiest way is to choose the music (files or playlists) from the Library and then right click on the selected tracks and select play. If you want to select multiple tracks just hold down the Shift button while clicking on the files.

The music will now start playing along with some rather funky visualisations. Opinion on these varies but you do have plenty of choice - in the basic installation there are eight schemes plus the option of simply displaying the album cover as a JPEG image file. You can access these choices by clicking on the little button on the top left hand of the screen with three small lines and a downward facing small green arrow. You can also call up further enhancements that include a graphic equaliser and add-ons like DSP sound processing systems.

You can also choose which skin your music will play in. The main screen is a little on the large side and for most people they'll be playing music in the background - a smaller, funkier skin is probably more useful. These come in all sorts of designs – sci-fi consoles come as standard but you can download a huge number of alternatives.

Media Player with sci-fi like skin

Ripping and burning using Media Player

Clever old Microsoft has done everything in its power to make sure that Media Player provides all the key tools that any serious digital music fan should ever require and that means it also boasts a number of rip and burn features.

Let's start with the ripping facility.

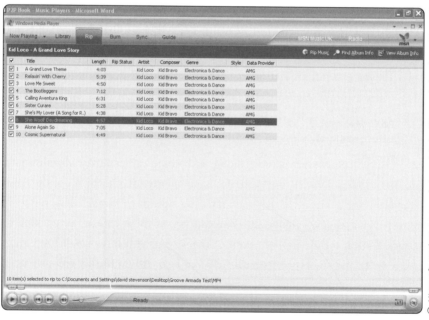

- Insert a CD in the drive and then open up Media Player.
- Click on the 'Rip' tab. You'll now see the music tracks listed, hopefully with all the music information. If the details of the album are not listed simply hit the Find Album Info button and Media Player will go online to one of the music databases and retrieve the information.
- You're now ready to rip the music. If this is the first time you've done this using Media Player you may see a new screen come up that gives you the option of adding copyright protection to the ripped tracks.
- Click on the 'Do Not Add Copy' protection and click on the 'I understand' box.
- Next you'll be asked it you want to use the pre-configured settings or use your own. Personally I'd select your own options.

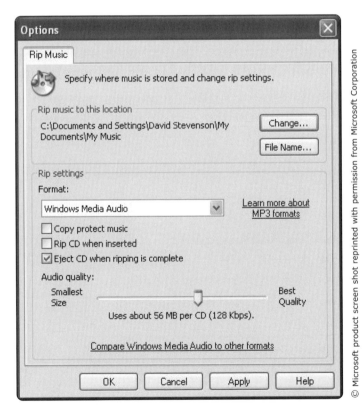

- If you do decide to set your own options you'll be presented with a number of different choices. First you can set-up where you rip your music too – in most cases this is best left as My Music within My Documents.

- You can also tell Media Player which codec to use for compression. Windows Media Audio is selected as standard. Personally, I'd change that to MP3 because it's the more widely used format and is guaranteed to play on nearly every portable music player.
- Click 'Apply' and then 'OK' and you'll be back at the main Rip screen. Now it's time to hit the 'Rip' button in the top right hand side of the screen. Media Player will now take a little over 30 seconds to rip each and every track. When it's finished it will eject the CD.

Burning music onto a CD

Media Player can also convert MP3 and WMA files back into WAV files that can be burned onto a CD for use in Hi-Fi systems (assuming you have a CD/DVD player that can burn discs).

- Click on the 'Burn' button on the main screen.

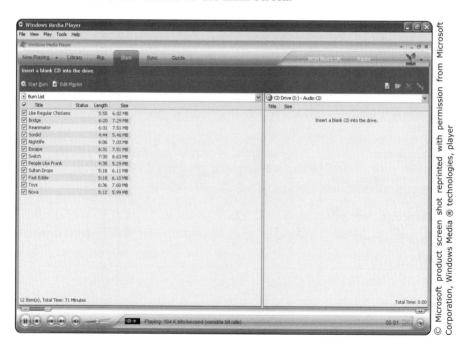

- Media Player suggests using playlists listed in its Library as a way of building up the files on the left hand side of the screen – the files to be burnt onto a CD. Frankly there's a much easier way – simply open up the folder in which your music is stored through My Computer and then drag and drop the files into the left hand – Burn List – box.

- Insert a blank CD in the drive and then hit 'Start Burn'. Media Player will now convert all the files to WAV format and then burn the CD. The conversion itself will take a couple of minutes while the burn will take five to ten minutes depending on the speed of your CD/DVD unit.

jetAudio

As an alternative to Microsoft Media Player, the pick of the bunch by far is a wonderful program called jetAudio (you can get it free at www.jetaudio.com). This enormously clever and stylish piece of software was originally designed by an electronics manufacturer called Cowon as a useful add-on for its MP3 music players (that are heavily advertised on its website). To get the software simply click on the 'Download' button and choose jetAudio Basic (there is a paid-for version that is very powerful, but completely unnecessary for most music aficionados).

Program profile – jetAudio	
User	Beginner
What it is	Easy to use music player, ripper, burner and converter
Why bother	It looks great, is simpler to use than Windows Media Player and has extra features
Source	www.jetaudio.com
Difficulty	Easy
How long will it take to master	A few minutes

Once you've installed the software you'll immediately notice that Jet is much smaller and more stylish than Media Player. It also incorporates a number of nifty player features like a graphic equaliser controller and it comes pre-loaded with extra tools that allow you to rip and burn a CD, plus conversion back and forward between different codecs.

The Downloader's Handbook

If you want to play some music you have a number of options. First, you can build up a music library by hitting the Album button. This launches a new box called Album Manager, which lets you build up the library of tracks that Jet draws from. To add multiple files simply click second folder symbol (the one with three folders stacked together) and then select the folders plus sub-folders you want to add to your collection. When you're complete, simply double click on the file(s) you want to play, and Jet will start up.

You can also select individual tracks or whole folders by clicking on the open files button on the main control pad. You'll now see the funky graphic equaliser screen power up – the music as a waveform. You can either choose to leave the levels set at Flat or select the music style that's most appropriate.

If you click on the Visual button (top right hand side above the time counter) you can also select visualisations to run alongside the music. My personal favourite is Pixel Trip, but you can add other visualisations from the Cowan/jetAudio website.

jetAudio boasts one very nice extra feature: an automatic fade up and down when playing new tracks. This is altogether a nicer way of playing tracks and comes as a default option in jetAudio.

jetAudio also comes with all the standard rip and burn features that are found in Windows Media Player.

First, let's access the rip tools. Insert a CD in your PC and then click on the Rip screen.

jetAudio – Ripping

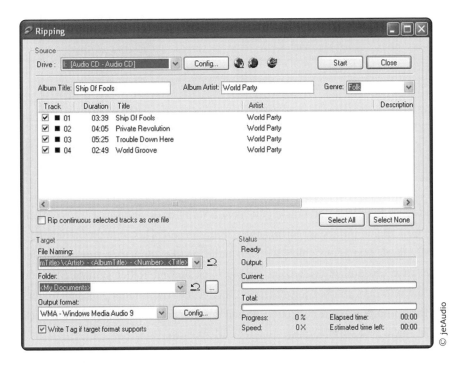

You'll now see the main Rip set-up box. Your key decision here is whether to rip the tracks as WMA files (the default), or OGG or even Musepack. You'll also notice that there's no MP3 choice available – as this is the free version of the software, Cowon/Jet can't afford to pay out for the license that allows them to use the MP3 codec, although the advanced paid-for version does have this option. When you're ready hit 'Start'.

The Burn tool is also very easy to use. Again, insert a blank CD you want to burn on, click on 'Burn' and you'll see a new box appear. You can now add the files you want to put on your CDs simply by using the Add Files box. When you're added all your files simply hit 'Start'.

jetAudio – Creating

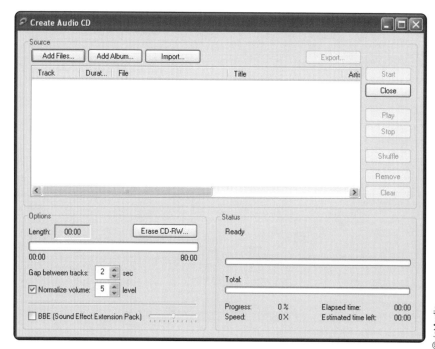

jetAudio boasts one other marvellous extra feature – a conversion tool that lets you move back and forward between most codecs, which is a great feature if you want to move between say WMA and the OGG format. To use it simply click on the 'Conversion' tab, select 'Convert' Audio and then find the tracks you want to convert from through the main Conversion window. Click on the 'Add Files' button and select some MP3 tracks, and then at the bottom of the screen select which codec you want to convert to – these range from Musepack through to WMA but again doesn't include MP3.

jetAudio – Converting

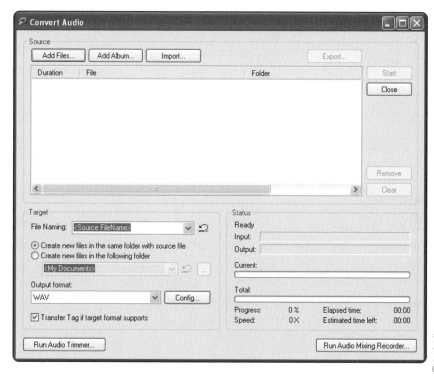

Winamp

The main alternative to Jet amongst experienced users is Winamp, a powerful but relatively easy to use program that lets you do everything Windows Media Player does and more. It's not quite as simple to use as Jet but it is backed up by a huge community of users who have devised all sorts of useful tools and add-ons (available at www.winamp.com/plugins/).

Program profile – Winamp	
User	Beginner
What it is	Easy to use music player
Why bother	It looks great, is simpler to use than Windows Media Player and it's also free
Source	www.winamp.com
Difficulty	Easy
How long will it take to master	Ten minutes

Playing music

First, run the program. A window will come up asking for details for registration. This isn't needed to use the program so check the 'Do not ask again until next install' box and click 'Later'.

Setting up a playlist

Now the program will open. To listen to a track you need to add it to the playlist. By default, the playlist window will be open but the windows have a habit of being accidentally closed. To open the playlist window, either click 'View' then 'Playlist Editor', click the button labelled 'PL' at the bottom right of the main window or press 'Alt+E'.

Winamp – Playlists

Click 'Add' in the window labelled Playlist Editor, then select 'Add file(s)' or press 'L'.

Now a browse window will open; from here you can select the file(s) that you want to open. Once selected, click 'Open'.

The file(s) will then be displayed in the Playlist Editor window. To play one, double click it. Alternatively, if you keep a selection of music that you want to play already in a single folder then you can add the entire contents by clicking 'Add' then 'Add Folder' or press 'Shift+L'. This will open a window from where you can browse for the folder to open.

To save a playlist, click 'Manage Playlist' then 'Save Playlist' or 'Ctrl+S'. To open a playlist, click 'Manage Playlist' then 'Open Playlist' or 'Ctrl+O'.

Media Library

It is also useful to set up the Media Library so you can listen to music without having to browse in windows, but using the tags of each track so you can search by artist, genre, etc...

The Downloader's Handbook

To open the Media Library window, click 'View' then 'Media Library', or click the button labelled 'ML' at the bottom right of the screen or click 'Ctrl+L'.

Click the Audio link under the Local Media on the left of the window. You will be prompted to add files to the library. Do this as though you were adding folders to a playlist. Once you've finished adding all your music click 'Close' on the pop-up.

Next you'll notice all your artists and albums are listed in the main part of the window. Select an artist or album that you want to listen to and the tracks will be displayed in the lower part of the window along with some web information about it.

Winamp – Media library

Double click a track to start it playing.

You can also use Winamp to listen to internet radio stations and even use it to download free music tracks. The internet radio music is done using a technology called streaming – we'll look at this in the next section.

The different lists can be found to the left of the Media Library window. Under Online Media is a choice of Shoutcast Radio, Shoutcast TV, Winamp Music, Winamp Video and AOL Video. The Winamp Music, Winamp Video and AOL Video give you a small variety of either music, music videos or TV snippets, designed to make you want to buy more. However, the Shoutcast radio and TV is entirely free content from an array of radio servers. To stream anything from any of the categories just select it and, after a couple of seconds, a list will appear in the main part of the window with all the available streams from that category. The first time you do this you will be asked for content ratings. Select the ratings that you want, check the 'Do not show me this again' box and click 'Close'.

Winamp – Radio streaming

Double click one of the streams, wait a few seconds for the buffer to fill up and enjoy free media.

If you ever get bored, press Ctrl+Shift+K, and sit back and enjoy a huge range of visualisations on your screen.

Portable MP3 players

I thought I should also include a mention of portable MP3 players, although this book is mainly about playing digital media on PCs.

Despite all the hype and excitement generated by Planet iPod, it's easy to forget the simple fact that only about 20% of the UK population has some kind of portable music player using either MP3 playback or Apple's variant based on MP4.

Which is a pity, because a portable MP3 player is fantastic – if you have the right one. Or two or even three!

Even if you have an MP3 player, what's to stop you buying another one? Maybe you could have a really small device (possibly flash based) for normal day to day stuff and a larger capacity one for travelling?

And if you're in the majority that hasn't even got one single MP3 player yet, you really need to get one – now!

Portable MP3 players are falling in price all the time and the technology is getting better and better; but do remember that not all MP3 players are created equal.

What to look for in an MP3 player

1. DRM – Digital Rights Management

    In the good old days, you downloaded music as an MP3 track and then copied it across to an MP3 player with no worries about clever software that controls the copyright. Those days are long gone. You need to take great care in working out if your music player has the appropriate rights protection or DRM. The biggest problem is that music purchased from the Apple iTunes music store won't play on anything but an iPod. So don't buy anything but an iPod if you're planning to shop at iTunes. Also, services like Napster restrict the number of music devices their music will play on – look for mention of DRM protection built in and check on Napster's website to see if your player is compatible. If it isn't, music purchased on Napster (or MSN Music for that matter) won't play on your MP3 player.

2. **Format**

   Most music players will play MP3 and WMA (the Microsoft format); while iPods will play the Apple MP4 format as well as MP3 and WMA. But the vast majority of MP3 players won't play Apple's formats. Beyond these hugely popular formats, it gets even trickier. A good player will also use OGG files – they're good quality, but the files are much, much bigger than MP4 or MP3. Do try to find a player with OGG support if at all possible. Also, the ability to play WMA though common is not global – make sure your player will play this format as well.

3. **Capacity**

   Most players fall into one of three major categories. Small capacity players will typically play up to 2GB with storage capacity starting at 64MB. 128MB is probably the absolute minimum (this will equate to an average of around 20 tracks at acceptable encoding rates); while 256MB is a happy medium (40 to 50 tracks). These smaller capacity players will tend to use flash memory, rather than PC-based hard drives – that makes them light and cheap, but it caps capacity at a few GBs. Mid-sized players start at 1GB and go up to about 6GB. They have enough capacity for hundreds of tracks and are also, usually, quite small. But they're also more expensive – you won't get much change from £100. The high capacity players start at 20GB and use proper PC-based hard drives, can carry thousands of tracks and are great for travel and holidays, but they can end up being a bit bulky. They also tend to be pretty expensive, with most decent models made by iRiver, Creative, Apple and Archos starting at £150.

4. **Power**

   Good players will give you at least 6 hours run time on either one small AAA battery (typical of the small capacity flash drives), or a single lithium ion recharge. Above 10 hours is great and below 3 hours is bad. Rechargeable lithium ion batteries are a great idea but tend to give you less hours of usage, while AAA batteries work out more expensive over the longer term but tend to feature in much cheaper devices that typically cost under £50.

5. **Screens**

   Manufacturers love to brag about their amazing all singing, all dancing, colour screens, but if you only play music, colour is not really necessary.

What is necessary is that the display gives you basic track information with at least three lines of data (enough for album, artist and track name).

6. Playlists

   These are a wonderful innovation and allow you to edit together your own 'Best Of' compilations. But most cheap MP3 players won't be able to handle these little text files (typical file formats end with M3U). They also tell the machine to play the music in a specified order, otherwise the player will revert to playing tracks starting with names beginning with 'A' or the number 1.

7. Weight

   I used to buy heavier jukebox like MP3 players (equivalent to iPods) but stopped recently. I found them heavy to lug around and expensive to replace. If, like me, you're a bit absent-minded, spending hundreds of pounds on a bulkier MP3 player with loads of features is overkill and a possible waste of money. Most of you will probably only want access to, say, 40-100 tracks on the move at the same time, which suggests buying a smaller (in weight terms) capacity player that is also much cheaper – ideally look for 1GB in capacity that costs less than £100 and also plays DRM music from the likes of Napster. Heavier players are wonderful if you travel extensively, but for most of us a light player will do. In fact, the smaller the better.

8. Music software

   The manufacturers of MP3 players love to brag about the amazing things you can do with your player and their special software. For example, they'll tell you about the fabulous possibilities of syncing tracks between your PC and MP3 player in nano seconds, courtesy of software provided by the likes of MusicMatch. Forget it. It's all a big waste of time, except if you've got an iPod. You should use your device as a simple, portable hard disk device. When it plugs into the PC you can drag and drop files simply by treating it as another hard drive by accessing it through the My Computer icon. It's all simple and easy using XP's Plug and Play capability. The big exception is the iPod where it does make sense to use its bundled software (you don't have a lot of choice, in truth) although you can still use the device as a simple plug-and-play hard drive.

9. Headphones

    Not all headphones are created equal, although nearly all headphones supplied by MP3 player manufacturers are (usually) equally dreadful. In my experience it's frequently worth paying less for the MP3 player and a bit more for the headphones. Good ones by the likes of Shure or Sennheiser are really very, very good.

10. Cost

    The sixty four thousand dollar question – the answer to which I've already hinted at earlier on. Personally, I wouldn't dream of spending more than £100 on a portable player in case I lost it. There are some great players above this price, but I'd be too worried about losing it (and lugging it about). You may be different, but think about this: for £50 you should be able to buy a perfectly good player that plays more than 50 tracks, weighs next to nothing and plays most main formats. If you lose it it'll be a pain but you should be able to afford another one in an emergency. If you insist on spending more than £100 make sure it's fully equipped with DRM, it handles as many formats as possible and has a good screen and decent battery life.

Internet radio and streaming audio

As we'll discover in this book the availability of digital music is by no means limited to your CD collection. There are, online, literally dozens of different networks that allow you to download music files over the internet – some of them entirely legally, some of them less so. These networks – like Napster for instance – involve some kind of download delivery system, which means that a piece of extra software sitting on your computer tells it where to find a music file and then download it to the right place on your computer.

But downloading isn't the only game in the digital music town. There's a widely used alternative called audio streaming. This involves a technology that is effectively a kind of internet radio. Instead of downloading a whole file, your computer receives small packets of data that are put back together by a piece of software in your computer as one continuous real-time stream.

This audio streaming technology is used to greatest affect by internet radio stations such as Live365 and the BBC.

live365

Live365.com is the mother of all internet radio stations. It acts as a central hub for literally thousands of free radio stations from all over the planet, that in turn play every conceivable style of music known.

Live365 – Front page

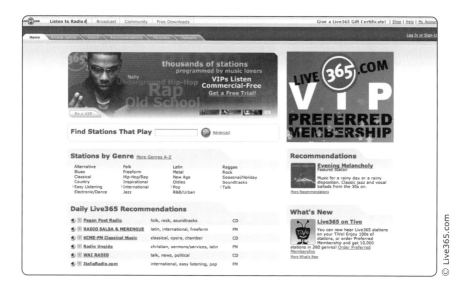

Live365, like many internet radio stations, uses its own software to play the music over the internet. Some networks have used third party programs, Microsoft's Media Player or the Real Networks player (we'll run into this with the BBC), but Live365 does away with these and has built into its website its own player that pops up as a separate screen when you click on a radio station.

listening to live365

To listen to internet radio, all you have to do is select the kind of music you want, and the station that best suits your musical tastes, then click on the 'Listen' button next to the radio station listing, and then the player window pops up.

As you've chosen the free version of Live365, you'll now have to put up with the highly annoying radio advert – these last about five to ten seconds – and then, hey presto, you're listening to your specialist Tibetan Goat Herding music station.

> **Tip**: The Live365 Professional stations are only available if you pay an annual or monthly subscription, but they boast better audio quality and no annoying ads.

BBC

The BBC also has an enormous wealth of audio content; both music based and spoken word (Radio 4 is especially strong on documentaries and current affairs). Thankfully, the BBC is one of the most technologically advanced media corporations on planet Earth and has moved heaven and earth to make sure that its most successful radio stations and programmes are available online.

The model here is a slightly different one in that it's actually closer to download technology – though some live radio shows are available in real time on the internet, the most popular programmes are actually library download copies that are available via a streaming network.

So far so wonderful – it's all free content and you don't have any annoying adverts.

The BBC downside

All the BBC's radio content uses a codec based on technology devised by Real Networks. Real was one of the pioneers of streaming technology and has successfully survived the onslaught of mighty Microsoft by developing proprietary player software that is still immensely popular with streaming radio stations. The only problem is that I see no earthly reason why you need any of Real's software. Its player is, in my experience at least, buggy and drains system resources and there are also loads of annoying ads. Apart from internet radio stations you really don't need any other tools built into its software.

But BBC Radio has, sadly, tied itself to Real and that means whenever you click on their Radio Player it will try and use the Real software that's supposed to be installed on your PC.

But there is an alternative!

Real Alternative

Its called, unsurprisingly, the Real Alternative. To download it simply type in the software name to any major search engine and you'll be taken to any one of the dozens of sites that host this free software. This gem doesn't install any of this Real Player nonsense on your computer, but instead uses a small program called Media Player Classic that installs all the necessary codecs without any of the spyware/system hungry player software.

Once you've installed this Real Alternative you can now click on any of BBC Radio's stations and you'll see its Radio Player pop-up as part of a web page. The BBC has also put a small number of its spoken word radio programmes into an online library. If you're into history and science for instance, the wonderful 'In Our Time' has made sure that all its past programmes are in an archive – available at:

www.bbc.co.uk/radio4/history/inourtime/inourtime_archive_home.shtml

If you like the look of a particular programme simply double click on it and you'll see the Media Player Classic screen appear. After a few seconds of buffering – this involves your computer contacting the host computer and downloading the start of a stream – the programme should play.

BBC Radio Player

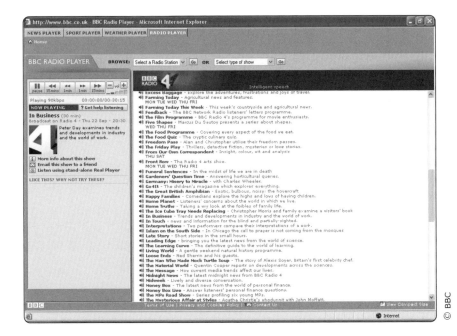

Recording music and speech using your PC

Internet radio is a wonderful innovation, but what happens if you want to record a favourite programme onto your PC in the same way you record your favourite TV programmes using a VCR?

Time to acquaint yourself with two rather wonderful programs. The first is a paid-for piece of software called Audio Recorder Pro, the second a free program that is a must for all music buffs called Audacity.

Audio Recorder Pro

Audio Recorder Pro is one of a number of software programs that allow you to record streaming music and speech through your computer. It's not necessarily the most high powered or even the cheapest of programs, but it works, is easy to use and comes with a number of very useful tools that are invaluable for recording audio streams.

Program profile – Audio Recorder Pro	
User	Beginner
What it is	A program that records audio coming through your PC sound card and turns it into an MP3 track
Why bother	Want to record your favourite internet radio stations? The one downside is that it's not free. It costs £15
Source	www.snapfiles.com/get/ezaudiorecorder.html
Difficulty	Easy
How long will it take to master	A few minutes at most

Download the trial Audio Recorder Pro software from its website at www.snapfiles.com/get/ezaudiorecorder.html. The paid-for version costs $25 (approx. £15) and is well worth the money.

Install as normal and then launch the program. You'll see a main window appear – before you start using it you need to make sure you've selected the right audio source. Click on the 'Source' box and you'll see a number of options appear, make sure you've selected 'Stereo Mix'.

Now it's time to find a radio station on the web. When you've selected your stream, head back to Audio Recorder Pro.

Audio Recorder Pro – Main screen

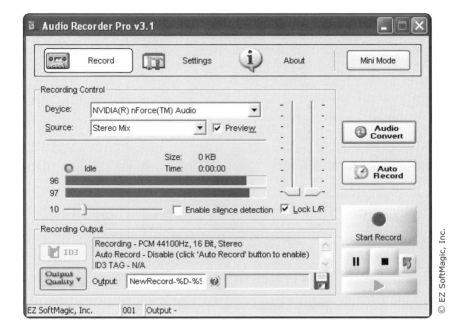

Let's record 45 minutes of a radio programme on to our PC

Audio Recorder Pro will turn the music that's coming out of your PC speakers into an MP3 track which can then be played back whenever you want – or can even be burnt onto a CD.

Click on 'Auto Record' and you'll see a new box appear which gives you the opportunity to run the program for a specified number of minutes (or hours). First make sure that the enable 'Auto Stop' function is ticked and then increase the 'recording by length' to 45 minutes. You can also use this box to schedule future recordings by using the Auto Start option. When you're ready hit 'OK' and go back to the main screen.

Audio Record pro – Auto start options

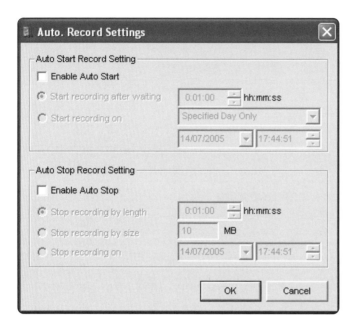

Now it's time to make sure that the MP3 track we're recording is of the right quality. Click on the 'Output Quality' button and a new box appears below the main window. Here you can control the settings for this recording session. Personally I use the MP3 format and the highest quality setting (HiFi) with a bitrate setting of at least 192 if not 320. This will produce a much bigger MP3 file but it will be of very high quality.

You also need to make sure that the source audio is not too noisy i.e. the audio levels aren't too high. On the main screen you'll see the two peak level indicators flashing first green and then red. If they're always red, your track will sound distorted. To lower the audio level inputs drag the two slider arrows down to an acceptable level.

After you've done this it's time to tell the program where we want to save the MP3 file to – look for a small Disk icon. Next to this you'll see another small box called Output. Type in the name of the file and then click on the Disc icon and save it to My Music (or wherever you save your music files).

You can, if you want, add an ID3 tag to the file for future media players to read. Click on the ID3 tag and you'll see another window appear where you can fill out all the boxes (remembering to tick the 'Write MP3 ID3 Tag' box).

Audio Record pro – Output quality management

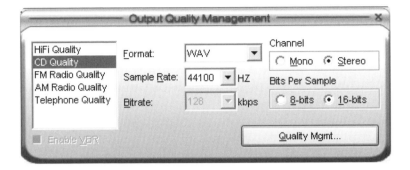

The Audacity alternative

There is a free alternative to Audio Recorder Pro called Audacity. It's important to understand that Audacity is not specifically designed for recording internet audio streams – it's a hugely powerful, all purpose audio mixing and recording program that's widely used by music buffs that can also be used to record safe audio streams.

Program profile – Audacity	
User	Experienced
What it is	A free music editing and mixing tool
Why bother	It can record sound and music from your PC sound card plus edit, mix and compile tracks
Source	audacity.sourceforge.net
Difficulty	Complicated
How long will it take to master	20-30 minutes

Once installed open up the main screen. Audacity isn't the easiest piece of software to understand but it's actually a lot easier than its main screen looks. Go to your internet radio station so that you can hear music or speech coming through your speakers.

Audacity – Main screen

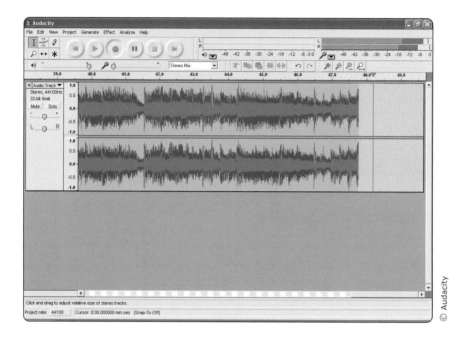

After you've installed the software open up the program – you'll see a big grey screen.

On the main toolbar (towards the top of the screen and below the buttons) you should see a box that says Stereo Mix. This allows you to set the audio input – in our case through the sound card.

Click on the small microphone button on the top right hand side of the screen. You should now see the audio levels go up and down with the music (or audio) coming through your speakers.

Now it's time to record this audio output. Click on the red 'Record' button. You'll now see a wave form diagram appear. Audacity is now taking in the audio signal and what you're seeing is the real time monitoring of what you're recording.

When you've finished recording the audio/music stream simply click on the yellow 'Stop' button. Now it's time to save the audio track in the right format on the right place on your hard drive. Click on 'File', and choose 'Export' as – you have the choice of either WAV, MP3 or OGG.

Having chosen one of these formats to export as, you're then prompted to choose the location of the file and give it a name. A five minute track should take no more than 20 seconds to export and save.

The downside – there's no timer to control the recording

Audacity, like many freeware programs, looks complicated at first, but is in fact blindingly simple to use. But this ease – and the fact that it's free – comes with one small problem for recording audio streams; namely you don't have a timer to control the recording.

To see why this might be a bit of disadvantage, consider a radio broadcast of 45 minutes. With Audacity you have to set the stream going and then hit Record and then remember to come back in 45 minutes and hit Stop! Now for many people that's not really too much of a problem, but for absent-minded souls (such as myself) it's a problem. Which is why I tend to use Audio Recorder Pro and its Timer features. Still, Audacity can do so much more than just record audio streams – you can also use it to edit music tracks and compile big mix compilations.

In this chapter we got to grips with ripping music from CDs, what codecs are all about, and what software to use to listen to digital music on your computer. We also looked at how to record internet radio and speech. Next up, where to find music tracks on the internet to download.

4

Online Music Services

4 – Online Music Services

It's time to put all this knowledge of music players and codecs to work, and head out online to see what music you can download using the internet.

In this chapter I'll first walk you through the two big UK-based services available online – the iTunes store and Napster. As you'll soon discover they both have their advantages and disadvantages, but they're both powerful online music libraries that should satisfy all but the most discerning of music fans.

But don't be fooled into thinking that your only choice is iTunes versus Napster.

There are some fantastic alternatives out there, not least from high street names like Virgin and HMV. And if alternative music like indie rock or jazz is your thing, you should also think about using the wonderful Wippit and eMusic. There's even a niche service that caters to audio book enthusiasts – it's called Audible and its online service has just launched in the UK.

But first you need to get your head around a terrible acronym that's becoming more and more of a problem for online music fans. It's called DRM and it's a powerful set of software-based tools used by the music industry to control the way you play and edit music purchased online.

You're about to discover that in this brave new world of legal, paid-for digital music you are not free to do whatever you want with your music!

In this chapter you'll learn how to:

- Use the iTunes music player and download music from the iTunes music store.
- Sign on and use the Napster service.
- Access the superb eMusic online service.
- Work with the Audible audio books service.

The world of DRM

Unless you've been living on a deserted tropical island with no access to world news, you've probably heard about all the rows over copyright and the internet. In a later chapter we'll talk in much greater detail about your legal position when it comes to downloading digital music and films, but for now let's make a couple of bold assertions:

1. Broadband and the growth of the internet has actually put you, the consumer, in a worse position as a user of copyrighted material. In the supposedly bad old days of analogue tapes and vinyl you had more power and more flexibility to do what you wanted with your music. Digital technology has actually reduced your flexibility to use this material.
2. Digital copyright technologies are not there to help you. They are there to hinder you and protect the music industry's ownership of their music.

Let's explain both of these overuse.

First I have to declare that I am no digital socialist. I passionately believe in the right of people who hold rights over a property to exploit them. Music is a piece of property, intellectual property nonetheless, just like any other traded thing. If you copy it and then give it to everyone else for free in perpetuity that is a form of theft.

But there is also something called *fair use*. If I buy a music CD then I should be able to do what I want with that music CD, as long as I'm using it for my own purposes. If I want to copy it to a tape or to another CD, or into a digital form which is then put on a hard disk and played by anyone in my household, that is not theft. It is simple fair use.

Except that according to many, but not all, in the music industry, it isn't.

With legitimate online digital services you only buy a license to use that music on the conditions set by the publisher. So if they say you can't play it on any portable MP3 player that they haven't agreed to, so be it. Your music is now disabled. If they don't want you to copy the music and convert it to whatever format you want, then you can't.

A very brief history of recording

To get inside the mindset of these intellectual property hawks cast your mind back to the bad/good old days (depending on how you look at it) of VHS. When the first VHS recorder was introduced the film industry fought it *every* step. They didn't want any technology in existence that could in any way copy anything that ever went on TV. Initially, they fought any copying facility, then they moderated their position by accepting that VHS manufacturers could design a machine that would only copy one programme/film once and then it would freeze up and stop playing. How useful!

Needless to say, VHS manufacturers were not too enthused by this restriction so they fought it through the courts. And won! Recordable VHS tapes were born and the film industry eventually made billions by flogging us loads of VHS films. With this mindset, you can see why the music industry reacted so badly to digital music – it put the fear of God in them!

But eventually they moderated their position and allowed us mere consumers to get music online through various music stores. But in doing so they built clever technologies that stop us from using our music freely (though there are exceptions like eMusic).

This practically means they can tell us how often we play our music, on what machines we play it and for how long and for what fee. The system that's used to do this is called Digital Rights Management.

The proper legal/technical definition of *Digital Rights Management* (or DRM) goes something like this:

> *"It's an umbrella term that refers to any of several technical methods used to control or restrict the use of digital media content on electronic devices with such technologies installed."*

The media most often restricted by DRM techniques include music, visual artwork and movies.

DRM effectively gives you no real legal right to fair use at all. It all depends on the largesse of the network owner and some are more generous than others. The actual tools used to implement DRM doesn't really need to bother you, but there are a number of practical implications.

Practical implications of DRM

1. Different digital music networks favour different music codecs. Microsoft loves its own WMA format but accepts MP3, along with Napster, but Apple only sells AAC based tracks through its iTunes store. Converting back and forth between these formats is not easy and in many cases technically illegal.

2. You do not have fair use to transfer your files how you want to. Most DRM schemes will try and restrict the computers you can use your music on and the portable devices you listen to your music on.

3. There is also a clear attempt to stop you using music if you no longer subscribe to certain services. Napster in the UK may only charge less than a tenner a month to access its huge library, but as soon as you stop paying, your access to your downloaded music on your PC vanishes.

4. Last, but (in my book) by no means least, this DRM effectively stops globalisation. Software has been built into computer servers on the internet that will recognise what country you're coming from – if you're from the UK that means you can't access US-based services even though they're cheaper. So you can import their CDs using various internet services but you can't use their music services.

If you can live with all of this – and most people can – there is one final point worth considering: cost. Why bother downloading? At the end of this chapter we'll compare the cost of the various download services with the good old fashioned high street retailers and a range of online CD stores. The results may surprise you.

The major music download services

msn

The vast majority of computers have Windows Media Player as standard and assuming you have internet access (preferably broadband), accessing Microsoft's own online music service is just a flew clicks away. But first, a few major cautions.

1. Media Player uses DRM. That means Microsoft has very substantial control over how you play its music.
2. It uses the WMA format to deliver its music online. That means you *cannot* use an iPod. iPods use a different DRM-controlled format called AAC.
3. Just because your player says it can use WMA doesn't mean it can use Microsoft's online music store. That's because DRM-backed WMA (sorry for the profusion of acronyms) requires more advanced players that can tell if a file is copyrighted or not.
4. Unlike some other services (Napster and eMusic) you don't pay for unlimited access to all the music in its library. You pay per track or album.

Service profile – MSN Music	
User	Beginner
What it is	MSN Music is easily accessed through the Windows Media Player
Why bother	Very easy to use through Media Player
Source	Incorporated into Media Player
Difficulty	Easy
How long will it take to master	Ten minutes

Assuming this is all fine and dandy with you (iPod users please skip this section) it's time to go online.

Open up Windows Media Player. Look to the right of the main screen and you'll see MSN Music UK. Click on this button and you'll be taken to MSN Music's main front page.

MSN – Front page

You'll now see the main rock and pop page with the latest releases. With each album or single you have three choices. First you can press the little loudspeaker/headphone button and get a thirty second (low volume) preview of the track. You can also hit the Play button and the track will be streamed to you – cost 1p per stream. (We'll talk about streaming a little later.) Lastly you can buy the track.

Buying a track

To buy the track click on the 'Buy' button. You'll now be asked to register for the service. You can now choose to pay upfront (using credits) for a number of options. You can either pay per track (69p the last time I looked) or £6.99 for an album. You can also bulk buy and pay £20 and get £22 credit, and so on (this pricing may have changed by the time you read this though!). Do note that this credit is only valid for 12 months.

Once you've signed up for the service (given yourself a user name and so on) you'll be asked to download Music Manager. This is designed by a software company OD2 (more on them later) and it manages the downloads and determines where the music is stored. Choose Install. You might also have to install an Internet Explorer ActiveX control. All in all this whole process shouldn't take more than a few minutes on a broadband connection.

Now, armed with an account, some money in that account and the Download Manager you can go back and buy the track you wanted. Click 'Buy' and you'll be taken to the payments page – let's pay 69p to download the track.

Once you have paid for the track go to the 'My Downloads' button on the main page.

MSN – My music page

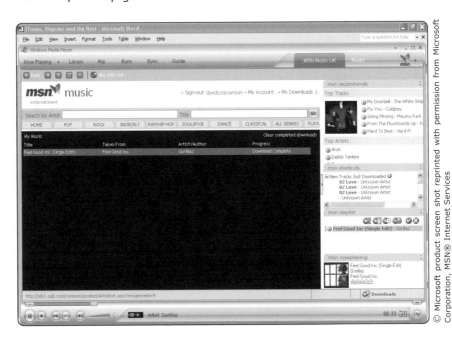

You may be prompted to download yet another piece of security/license software – click 'Yes'. Now the download starts. On a 512Kbps or 1Mbps connection it should take no more than a couple of minutes.

Once the download is complete you should see it in the My Downloads/My Music list. Hit the 'Play' button at the bottom to play the music.

Syncing your portable MP3 player

You can also transfer this newly purchased music to your portable music player. Once Music Manager is installed, simply connect your music player and you'll be presented with the option of 'Configuring the sync' with your portable device. Install this. Then go to the Sync button on the main Media Player screen.

From here you can control how you sync your music with your device. Connect your device to its USB lead, and Plug and Play will see the device. On the Sync page you will now see all the files inside your device on the right hand side of the screen.

Windows Media Player – Synchronising MP3 players

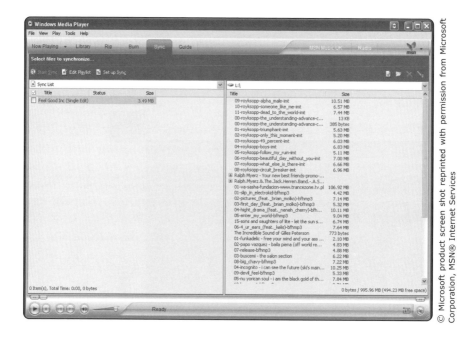

Now it's time to set-up how you sync your device. Your first option is to do it automatically. Press the 'Set Up Sync' button on the top left hand side. This gives you the opportunity to automatically sync any folder inside Media Player's Library. Click any box you want to automatically sync up every time you connect your device.

Synchronisation settings

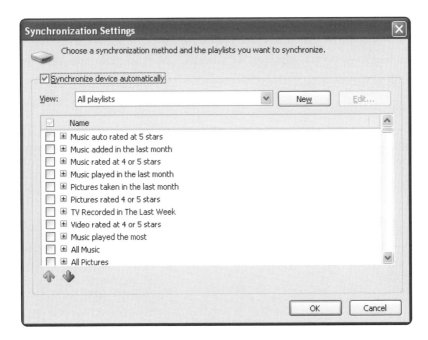

You can, as an alternative, choose simply to select your own files to synchronise – click on 'Edit Playlist'. You can now navigate your way through your music library choosing whatever tracks you want.

Warning: If your portable music player doesn't support DRM, you'll soon encounter a problem with this sync procedure and your downloaded (purchased) music files. They won't sync with your machine!

Burning your purchased music to a CD

You may decide that you'd rather keep your music on a CD for use on your Hi-Fi system. You could, for instance, take your burnt CD and then rip it again and turn your music into copyright-free MP3 tracks, which you can then play on whatever device/system on earth you choose to!

MSN Music UK

If, like me, you decide you like the sound of burning your music to your CD simply go back to the MSN Music UK page and click on 'My Downloads'. Now look at the bottom right hand side and you should be able to see MSN Playlist as an option.

You'll see six icons, one of which is an icon for a CD. Once you've selected the music from your downloads that you want to burn hit this button. A 'Queue List' box now appears at the bottom right of your main screen (an MSN Music icon should also have appeared in your taskbar list).

Your main Media Player screen has now gone to the Burn page. You should now see your chosen track/tracks listed on the left hand column. As you have a blank disc in the burner, your right hand column (the burner) should be clear.

Windows Media Player – Burning

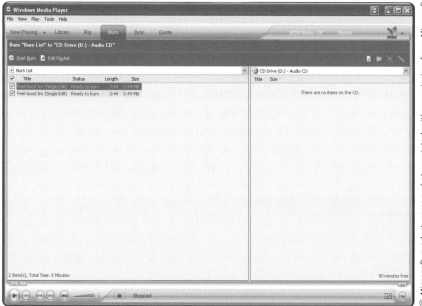

Once you've lined up all the tracks you want to burn onto CD, you simply press Start Burn and Media Player handles everything else. It converts the files from DRM-protected WMA to WAV and then starts the burn. A few minutes later your CD is ready and waiting.

You could of course now place the CD straight back in the tray and rip the contents of the CD to copyright free MP3 and do whatever you want with the tracks – not that Microsoft would approve!

In fact, you'd be annoying more than Microsoft – you'd also be annoying the company behind this and many other services in the UK, OD2. This outfit runs the backend of a number of digital music stores – they all share the same technology and software support.

These digital music stores include:

- MyCokeMusic
- Wanadoo
- Tiscali Music Club
- HMV Digital Downloads
- Virgin Downloads
- Ministry of Sound
- And, of course, MSN Music itself

The various OD2 services

The only major difference between any of these services is the cost.

- MSN Music is usually the cheapest, with charges pegged at 69p a track and £6.99 per album.
- MyCokeMusic charges 99p a track, or £7.99 an album.
- Wanadoo does offer a special service that allows unlimited streaming for just £4 a month. Otherwise it's 79p a track or £6.99 an album.
- Tiscali also charges 69p a track and £6.99 an album.
- Virgin, like MyCokeMusic, charges an expensive £7.99 an album or 99p a track; although if you bulk buy credits this does come down to a mere 80p a track (still not cheap!).
- Ministry of Sound is great for dance music, but pricey at 99p a track or £7.99 an album.

All these prices were correct at the time of writing and may now have changed.

Do you own this music permanently?

Yes and no. The Ministry of Sound helpfully spells out precisely what rights you're buying into with the OD2 service.

Here, in their words, is their definition of your *permanent rights*.

> "Number of permitted file downloads = 5. This means that you can download the track up to five times after you have purchased it. This is useful if you want to have one copy on your computer at home and another copy on your computer at work for example. This also allows you to download the file again if you accidentally delete it from your hard drive or you replace your computer.
>
> Track can be copied to portable device = Yes.
>
> You can copy your download to a portable MP3 player an unlimited number of times.
>
> Number of times the track can be burned to CD = 5.
>
> You can burn your download to CD up to five times. If you want to burn the download to CD a sixth time you will need to purchase it again.
>
> Downloads are only available in certain countries due to music licensing regulations (UK and Ireland)."

So, yes, you own the music; but then again, no, you don't. Remember the good old days when you bought some music – end of story. This DRM and licensing regime is restrictive and frankly the only way of breaking free of it is to pay for the music, burn it onto CD and then use it however you want to by ripping the CD and converting everything back into the MP3 format.

Apple iTunes

Before we delve a little deeper into the only real challenger to Microsoft in the music player and digital download market, lets get a few things clear from the start.

- Despite Apple's audacious marketing and hype about revolutionary new services, iTunes is almost as restrictive in its rights policy as OD2.

- Its music store uses a form of DRM that requires you to use the AAC codec, itself a derivation of something we've already encountered called MP4. This is both good news – on iPods this codec sounds great and it's a very efficient compression technology, and also bad news – you can only play AAC files on iPods.
- The iTunes player is truly versatile and in its PC version is a real challenger to Media Player. But it's nowhere near as revolutionary as it likes to claim. It's good, but not that good. If you don't have an iPod and you have access to Media Player I'm not really convinced that iTunes is worth the extra bother.
- It's worth mentioning that iTunes is not an all purpose media player like Windows Media Player, – it doesn't directly play compressed DivX or XviD videos (yet), although you can access this facility through the Apple Quicktime software that is bundled with the main installer.

So why bother dealing with what iTunes has to offer?

The bottom line is that iTunes as a piece of software is excellent and the music store built into the software is, by and large, wonderful.

Service profile – iTunes	
User	Beginner
What it is	The iTunes player is a full multimedia player, and a way of accessing the iTunes music store and it's designed to work with iPods
Why bother	Great alternative to Windows Media Player
Source	www.itunes.com
Difficulty	Easy
How long will it take to master	Ten minutes

Use iTunes if any of the points below matter a lot to you:

1. If you've got an iPod you have to use iTunes. You have no choice.
2. If you're after one of the biggest ranges of music in an online music store, iTunes is for you.

3. If you're after a sleek interface and clever design that is much better to use than any rival player, with the exception of jetAudio.
4. If you like audio books delivered through the iTunes music store.

In practice, it will do you no great harm to load iTunes anyway, even if you continue using rival players.

Setting up iTunes

Tip 1 – Make sure you get the latest version

Whatever you do make sure that you're installing the very latest version of this software. You may have been given a version of iTunes with an iPod and that may be months, even years out of date. Apple regularly updates its software and it is worth your while making sure you have the latest version – at the moment it's 6.0.4 but it could change any time now.

Tip 2 – Get the UK version

There is a US version of both the program and the music store – but you have no choice other than to use the UK version. If you visit the main Apple.com website you'll spend a large amount of time fannying around with the US version before it eventually suggests getting a UK installation.

Make your life easy. Go to www.google.co.uk and type in 'itunes' and then hit the main link to their UK sub site – at www.apple.com/uk/itunes/. Click on the 'Download' button and then choose the Windows version.

- Install the program as normal.
- During this process you'll be asked on one screen to create a desktop icon (click 'Yes') and make iTunes the default player for your audio files. Personally I don't check this as I use Windows Media Player as well. You'll also be asked to make an associated program, called QuickTime, the main player for media files. Do not click Yes to this – there are better default media players out there and for streaming of video content you're better off using something called QuickTime Alternative (more on this in the next chapter on videos).
- Eventually the program will finish installing and you'll notice two new icons on your desktop: one for iTunes and one for QuickTime.

- Time to start iTunes for the first time. If you're online your firewall will almost immediately notice iTunes trying to connect to the web – let it. You'll then have to agree to Apple's license agreement and after that you'll run into iTunes' set-up screen. This will ask you if it's OK to search your computer for media files and then put them in its media library. That's fine, it's a great idea and won't affect any other programs you have running, like Media Player.

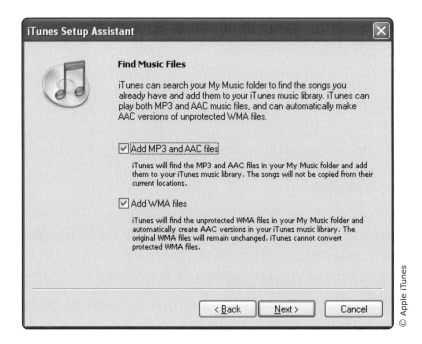

- Next click 'Yes' on letting iTunes organise its folders and then you'll be asked to go to iTunes' music store where you can register and join up. It's up to you if you want to do this – you can always do it later – but for now just say 'Yes'. You'll now land on iTunes' main page and be taken to the iTunes' music store. Almost instantly you'll be confronted by Apple's slightly relentless bullishness – yes, that's 500 million tracks downloaded from its store!!!

- You now need to sign up for the service. In the top right hand side of the screen you'll see a little button that says 'Sign In'. Click this. You can now create your own account (as you would with any online music store) or sign in if you happen to already have an account.

iTunes – Music Store

© Apple iTunes

- It's important not to make a mistake here, as the screen to sign up for a new account actually takes you, by default, to the US page. You can't join the US service – which is a pity as the songs are cheaper! You will see an option to 'click here' if your credit card is registered outside the US. Click on this. You'll now find yourself at the UK home store page. Click on 'Sign In' again and create a new user account. You have to sign up for their license agreement and give your credit card details. Eventually you'll pass through all the hoops and get signed in. You're ready to go.

iTunes – Creating a new account

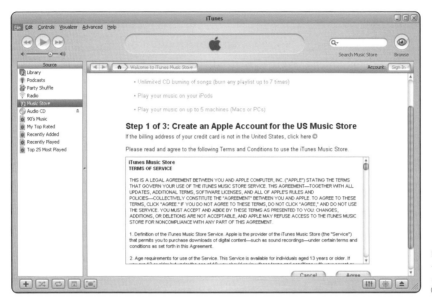

Beware of the first new user screen – it's only for the US!

Tip – get rid of QuickTime

I'm going to be a bit controversial here and suggest you uninstall QuickTime. Once upon a time QuickTime was everyone's favourite video program – not anymore. Times have changed and there are a number of rival formats that are, frankly, more widely used – namely Real's various formats and Microsoft's media files. 95% of the time you won't need Quicktime and if you do there is an alternative called QuickTime Alternative. Also QuickTime has a nasty habit of always loading up on start up (you can stop that) and of being generally a bit pesky and interfering. In my experience you don't need it, so get rid of it.

Some iTunes basics

iTunes – Library

© Apple iTunes

First, take a look at the layout. On the left hand side of the screen you'll notice a list of files and options. Let's start with the Library.

If there are no files in the Library you need to add them. Click on the 'File' option at the top left hand side and then select 'Add Folders to Library'. Most people will have their music inside My Documents, but if not you can tell iTunes where to look. If, like me, you have an awful lot of files, expect this process to take some time. iTunes will also convert any WMA files to AAC (the Apple format).

You'll now see a long list of files organised into different columns. The first column (starting on the left) is the track name, followed by the duration of the track, the artist, the album title, genre and then My Rating. These are pretty self- explanatory – the important bit is how to display the list so that you can make sense of it. Personally, I use the album as the way of organising my Library – click on the album column at the top and it will automatically order all the content into albums – starting with albums beginning with an 'A'.

When you've found the track you're looking for, double click it – the player will now start. If you want some fancy visualisations while you're listening to the music click on the middle button in the bottom right hand corner – the symbol looks a bit like a representation of an atom. You'll now see a randomly selected visualisation appear. If you want to turn this visualisation off simply click on this icon again.

Back on the main library screen you can also edit information relating to this music file. Simply right click on the file and you'll see a number of options. 'Get Info' retrieves all the ID3 tag information that you can now edit. You can also rate the track, which is a handy feature. If you award, say, five stars to your very favourite tracks you can then organise your library so that it plays the five star tracks first (simply click on the top of the My Rating column and it will sort all the files starting with five stars first). Alternatively you can select these files by looking at the left hand column and selecting 'My Top Rated'.

You can also change the texture and feel of the music played by using the graphic equaliser. This is accessed by clicking on the button next to the Visualisation icon – it's an icon with three channel mixers. Have a play with the equaliser and see which settings suit your speakers best.

4 – Online Music Services

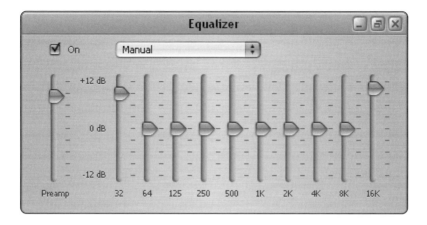

Another nice feature of iTunes is its Party Shuffle. This randomly shuffles up all your tracks and then creates a long playlist which you can use at a party.

iTunes – Party Shuffle

© Apple iTunes

You'll soon notice that the elongated grey-green bubble at the top of the screen is the 'progress' box. This shows you how long the track is – simply drag the slider along if you want to fast forward through the track.

Ripping tracks

You can also use iTunes to rip music from a CD and import it into your music library. Simply insert the CD, and run iTunes. It now searches one of the online music databases and retrieves all the track information. Before you click on Import in the top right hand side of the screen, check your ripping options. Go to 'Edit', and then select 'Preferences'. This box controls how iTunes operates and you'll see a number of options that include how your iPod connects and one that says Importing. Click on this tab and you'll see that you can configure iTunes on how to rip the file. It will by default use its own AAC format, but you can tell it to use MP3 if you want. You can also set the quality of the encoding: high is the default at 160Kbps and for most rips that should be sufficient.

While we're in this Preferences Box, click on the 'Advanced' option. Here you'll see where iTunes stores ripped and downloaded tracks. As a default it creates a folder inside My Music (which is in your My Documents folder). You can if you want, change this to whatever folder you want.

While iTunes rips the files it'll also play you the music – a nice little feature.

iTunes preferences

One last feature worth noting: you can share your music across a network. Let's say you have a bunch of computers in your house, all of which are sharing a wireless connection. Using the Preferences box, click on 'Sharing' and enable 'Share My Music'. Hey presto – you can stream music across the computers in the network.

iTunes' great virtue is ease of use. It really is as simple as this to use – build a library and then start playing. But this is only the beginning, because the player also seamlessly integrates with the online music service.

Buying and playing music from the music store

Let's buy some music from iTunes.

Click on the 'Music Store' green icon.

You now see the store front page. You can immediately click on any of the featured albums and then choose to buy either tracks (79p) or albums (prices vary but most major albums are £7.99). To do this go to the album page and either click on the 'Buy Album' or 'Buy Song' button next to each track listing.

Like all its competitors you can get a thirty second preview of the track. Simply go to the album page and double click on the track name.

iTunes – Music Store

To navigate around the music store you either use the backwards and forwards arrow buttons towards the top right side of the main frame, or if you want to go back to the home page simply click on the button featuring a small house symbol.

Once you've decided that you want to buy some music (assuming you're registered as a user with a valid credit card), simply click on the 'Buy Album' or 'Buy Song' button. iTunes will check that it's OK to charge the track/album to your credit card and within a few seconds you'll see the progress box (the grey-green long bubble) download the track into your iTunes' folder.

To play downloaded music simply look at the main left column (the one that starts 'Library') and you should see a folder called Purchased Music. Click on this box and you should see your track.

If you then want to burn this track/album to a CD, simply insert a blank disc, choose what tracks to put on the disc (by ticking or unticking the little box next to each track) and then click on the Burn Disk icon in the top right hand side of the screen. iTunes will now go away and burn the disk.

iMix

The iMix tool is a rather cute little feature of iTunes. The idea behind this is simple: you compile a list of your favourite tracks or maybe even a mix, and then either share the playlist with your friends (who must also have iTunes) or publish it to the iTunes website. Here's how you compile your own iMix:

1. First, you'll have to create a playlist from scratch. The easiest way to do this is go to a Library. Hold down the Ctrl key and click on all the tracks you want to include in a playlist. Then go to File (at the top left hand of the screen) and New Playlist From Selection.
2. You now see a new playlist file emerge in the left hand column. Next, give the playlist a name.
3. Now select the playlist you want to publish as an iMix.
4. Click the Publish arrow to the right of the playlist file. If you don't see the Publish arrow in the Source list, make sure the 'Show links to Music Store' checkbox is selected in the 'General' pane of iTunes' Preferences.
5. Enter the requested information about your playlist and click 'Publish'.

These iMixes include 30-second previews of any songs in your playlist that are also available on the iTunes Music Store. But it's also important to note that if a song in your playlist is not available on the Music Store, the song won't be listed in the iMix. Also, changes you make to your playlist after it's published are not updated automatically on the iTunes Music Store. You need to click the Publish arrow each time you make a change that you want to share.

You can also buy other people's iMix list. This is actually quite a good way of buying a compilation of what other people think is great music in your genre – an easy way of finding the best compilations is as follows:

- Go back to the Music Store.
- Click on the 'iMix' option in the left hand column.
- You now see a whole range of iMix's come up.
- Type the genre or bands you like in the search box – a shortlist of iMixes is now displayed. Look at the right hand side of the page and you'll see 'Sorted By Name'. This means the chosen iMixes have been chosen by name. Click on this box and choose 'Sorted By Top Rated' instead – this puts the iMixes that ordinary punters have highly rated first.

iMix – Music Store

When you see an iMix you like double click on it and the playlist will appear. You'll also see the total number of songs and how much it would cost you if you had to buy all these tracks (if you do want to do this click on Buy All Tracks' button next to the price).

Celebrity Playlists

An alternative take on publishing your playlist is to see what music your favourite artists like – this is the idea behind Celebrity Playlists. You can access this by again going to the main Music Store and going to the bottom left hand side of the page.

iMix – Celebrity Playlists

You'll now see a rogues gallery of famous (and not so famous) faces. If one takes your fancy just double click on the photo.

You'll now see the Celebrity Playlist in detail. First, in the main screen at the top you'll see their comments on the music. You can also buy all the tracks in the playlist plus you can go to the main playlist and hear a 30-second preview of all the tracks in the list.

Podcasting

Many people think that podcasting is yet another Apple innovation alongside its hugely successful iPod player. Far from it – podcasting is a simple audio-based diary, or blog, which has been transformed into an MP3 track. In effect, it's just a personalised radio station that you host at a blogging site, or at your own homepage. But Apple have been very quick to spot the potential for podcasts and built in a facility to download your latest favourite podcasts using iTunes. You can in fact subscribe to podcasts through iTunes and each podcast will automatically update inside iTunes every time the broadcast is changed.

The quality of podcasts varies enormously; there are some amazing shows that belong on proper radio stations you can access for free and some utterly terrible services that should be avoided at all costs. The point is to have a hunt around, looking for topics that interest you.

To get podcasts using iTunes go back to the main page. Click on the Music Store folder in the left column. You'll now be inside the main Music Store page – to the left you'll see Podcasts. Click on this button.

iTunes – Podcasts

You're now on the main Podcasts page. Here you'll see highlighted some of the more prominent podcasts available. You'll rapidly discover that most of the best podcasts are actually public service radio programmes repackaged as podcasts.

If you see a show that takes your fancy on this main page simply double click on the advert and you'll see more details about the podcast. You'll also see how many podcasts have been broadcast in the past and get the chance to download these as well. If you want to subscribe simply hit the Subscribe button and every subsequent podcast of the show will be downloaded to your iTunes library.

iTunes – Podcast Description

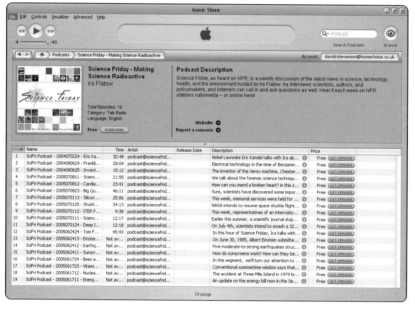

This first podcast subscription now means you'll see a Podcast symbol appear in the main left column on the main iTunes page. Double click on Podcasts and you'll see all the services you've subscribed to – after the latest episode has been downloaded all you need to do is double click on the episode in the main playlist.

If you want even more podcasts you can go back to the main podcasts page at the music store and search for appropriate broadcasts by category or theme.

You'll see a search option in the middle of the page on the left – you can enter a subject or you can click on one of the main subject areas below it.

You'll now see the main Search page. Simply select your main category and then your sub-category and you'll see a full list of podcasts available. If you like what you see simply hit the 'Subscribe' button and iTunes will automatically sign you up and then update the broadcasts.

iTunes – Podcast subscription

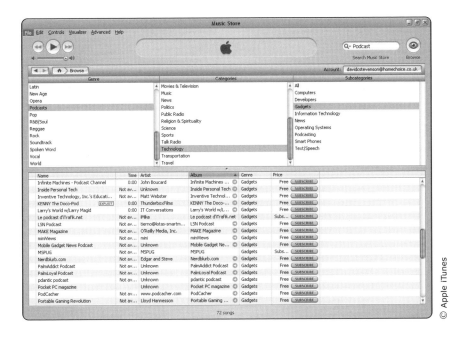

Radio

We looked at internet radio in the previous chapter and, like many other players, iTunes supports listening to internet stations from within the main player.

- Go to the main iTunes screen (the one marked 'Source' at the top) and double click on the Radio icon.

iTunes – Radio

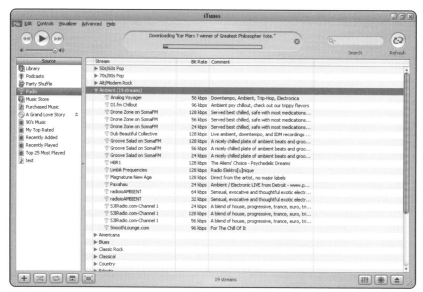

- You'll now see a large number of music genres appear, ranging from 50s/60s pop to urban. If a particular genre takes your fancy click on the Play arrow button next to the genre and you'll see a list of all the stations that are broadcast playing this type of music.

- If you click Ambient for instance you'll see 19 options listed. They range from 'Analog Voyager' which promises 'Downtempo ambient' through to 'Smooth Lounge, For the Chill of it...'

- With some stations you'll also see a number of different options: they're the same programming but they are broadcast using different quality settings. These quality settings (called *bitrate*) might range from a rather poor 28Kbps through to high quality 128Kbps or even higher.

- If you like what you see just double click on the station (preferably at the highest bitrate setting) and it will start playing.

Napster

You may remember Napster from its earlier incarnation as the poster child of the underground file sharing movement. This was the first great peer-to-peer network where you could download hundreds of thousand of different MP3 tracks, all for free. Ah, the good old days! And then the massive media combines ganged up on it and it was forced to shut down.

And then a strange thing happened. Napster turned from hunter to gamekeeper. It was sold off, to German media giant Bertelsmann, and relaunched first in the US and then in the UK as a legal music download service.

Its main appeal is that it offers a flat rate fee service. You pay £9.99 a month for unlimited access to all the music in its vast library. That means you can download any of its 2 million tracks. But be warned: these streamed tracks can only be played on your PC. If you want to play your Napster tracks on your MP3 player you must have a compatible MP3 player. The range of music is bigger than with MSN Music/OD2 and iTunes but by no means is every type of music included.

You can also choose to burn your chosen music files to a compact disc but you have to pay additionally for the privilege. This is charged like iTunes and MSN Music on a per track or per album basis.

One final fly in the ointment: stop paying the monthly subscription and your music will stop playing if you've bought it through the subscription service. Start the subscription up again and your music selection restarts.

Tip: The bottom line is that Napster is great for people who spend inordinate amounts of time sitting in front of their PC. If you use iPods a lot, forget it! iPods will not accept Napster's tracks.

Service profile – Napster	
User	Beginner
What it is	Napster's music service plus its music player lets you access 2 million tracks online
Why bother	Great alternative to the iTunes music store
Source	www.napster.co.uk
Difficulty	Easy
How long will it take to master	Ten minutes

Getting started

If you do decide to give Napster a go (with a 7 day free trial) go to www.napster.co.uk and download the main music player/download software. Install it as normal and then run the program.

- Napster will ask you first to sign up for an account and provide it with a credit card to make the monthly charge.
- Once you've got a sign in name and set a password Napster will communicate with the main server and take you straight to the main home page.
- You can use Napster as your main music player (replacing Media Player) if you want. Simply go to the blue Library tab at the top and click on 'File' at the top of the page. You'll now see a drag down list – choose 'Import Tracks to My Library'.

Napster – Homepage

- You'll be asked to select all the folders you want to add. Remember to keep 'Search Subfolders' ticked. On the right side of the box, under Files, you'll see all the WMA and MP3 tracks. If you want them all added simply press the 'Select All' button at the bottom of this list of files and then 'OK'. After a few minutes all the files will be added to the main Napster folder.

- You'll see a long list of compatible music files listed in the main Library window.

- This database of music files is like any of the other music players. If you want to order them simply click on the column header.
- If you now want to play a track, double click on the file you want and you should see the track plus any album art that's available appear in the right hand 'Now Playing' column.

Radio

Like all the major players – iTunes and Windows Media Player – you also have access to online radio stations. Look at the right 'Now Playing' column and in the middle you'll see a radio button. Click and then drag down the various stations available. The list is not as exhaustive as iTunes or Media Player but in my experience the stations are high quality. Once you've selected a channel Napster loads up the audio stream and you can now see the track details listed in the box above. If you don't like the track – you'll see the full list of tracks in the box – you can simply skip to the next track in the list by double clicking. The bottom line is that Napster's radio service is very cool and appreciably better than the competition.

Downloading music using Napster

- At the top of the screen you'll see a search box. Type in a band whose music you want to download.
- When you've located the track in the main search screen you'll see details about the album it's from. From this page you can choose to buy the album/track (for CD burning) or you can download the track.

4 – Online Music Services

- Right click on the track listing (if you hold Ctrl down and click you can highlight multiple tracks at once) and then select 'Download'. A few minutes later the track will have downloaded.

- Now go back to the blue Library button at the top of the page. On the right hand folder bar you'll see a folder called Download Status.

- In this list you should see your downloaded track with 'Complete' marked next to it.
- You can now play this track however many times you want – providing you keep that Napster subscription going!
- You may want to purchase individual tracks or albums for CD burning or transfer to non-compatible MP3 players. Again, search for the tracks/artist and when you find the track you want simply right click on it again and this time choose 'Buy Track' instead. A new box will appear asking you to confirm your password and then confirm the purchase.
- The track will then download for a few minutes and when it's complete it will appear in the Purchased Tracks folder back in the Library. You can now choose to copy the track onto a blank CD or onto a non-compatible MP3 player.

eMusic

You might have spotted a rather cautious tone creeping into my descriptions of the major online digital music download services. They're a huge improvement on nothing, but they're still a little cumbersome, rather expensive (we'll talk about this in more detail later), and they have pretty restrictive digital rights management systems built into them, which controls what you can and cannot do with *your* music.

There are no such problems with pioneering eMusic – you can find it at www.emusic.com. It was one of the first legal music download services – it started way back in 1998 – and is still one of the best.

Why is eMusic so good?

- First, it's cheap. The basic service costing $9.99 a month, lets you download up to 40 tracks a month – that's just over £6 a month. Even the premium service, which allows 90 downloads a month, is just $19.99 or just over £12 a month.
- Any music you buy is yours – end of story. You can burn it, send it to another computer in your home network or play it on whatever music player you want. There's none of this fierce copyright protection that spoils the Napster (and Apple iTunes) offering.

- And last but by no means least, eMusic has a reservoir of music (70,000 albums and hundreds of thousands of individual tracks) you're unlikely to find at the major music sites – it's a 'serious' music fans' nirvana, full of alternative rock, jazz and folk.

> **Tip**: If your music tastes are very mainstream, and Coldplay represents your idea of alternative, you're likely to find eMusic a bit strange. It's not jam packed full of major record label releases and although you will find the odd few tracks from the major labels, it's resolutely, even defiantly, alternative.

Service profile – eMusic	
User	Beginner
What it is	eMusic is an eclectic music download service
Why bother	Great alternative to iTunes and Napster for music enthusiasts
Source	www.emusic.com
Difficulty	Easy
How long will it take to master	A few minutes

Signing up to eMusic

If you do decide that eMusic is for you, sign up for its free trial at www.emusic.com/promo.html. You'll get 25 free music downloads for nothing. But be careful:

- The registration screen will sign you up and take your credit card details and unless you cancel, it will automatically sign you up for the more expensive Premium service when you finish downloading your 25 free tracks. If you only want the trial and nothing else be sure to cancel!
- After you've set up your account, you can browse online through eMusic's web pages. Once you've found an album you like – it's very much an album-orientated service – you simply click on the link and you'll be taken to the album page.

eMusic – Free Trial

- Like all the other music services you can either preview the track (click on the small loudspeaker icon), or download it (the down arrow icon).

> **Tip**: The best way of downloading music is get the eMusic Download Manager – you can get this at www.emusic.com/dlm/download.html. Follow the instructions and download the program. You can instead choose not to use the Download Manager and simply click on the music file you want and download it conventionally as a simple file – go to your account page, look on the right hand side and click on Download Options. You'll then be given the change to either Disable eMusic download manager if you've already got it and you can tell eMusic to save any downloaded tracks as proper MP3 tracks (otherwise it will save them as rather mysterious .emp files which won't play in any other media player). If you do decide to use the Download Manager, all your downloads will be handled by a dedicated piece of software – much the preferable option.

eMusic – Download Options

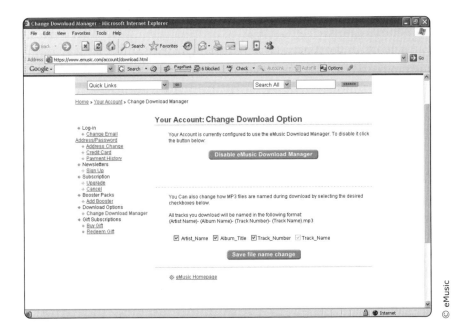

Once you've installed the program, simply browse through the eMusic web pages and find an album you want. When you've found one, click on either the individual tracks or the 'Download All' button.

Once you've done this the Download Manager starts up.

eMusic – Download Manager

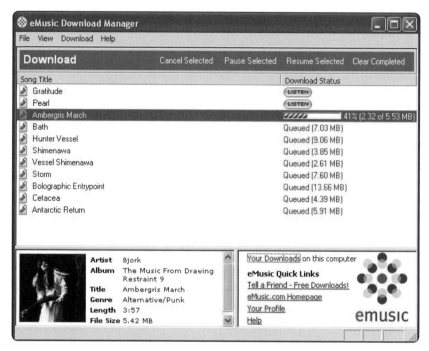

The Download Manager controls all your connections and you should customise the various options as soon as you start using it. The most important thing is to tell the program where you want your music saved to – click on 'View', select 'Options'. You'll now see a box called General and you'll also see a location where your music will be saved to. The default is a My Music folder on your desktop. I'd change this and tell Download Manager to save it to your main My Music folder inside My Documents.

eMusic – Downloading music

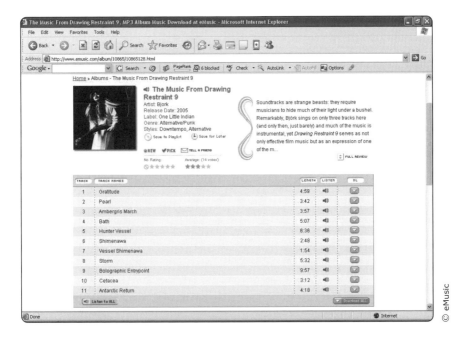

You'll also see all the music tracks downloading – over a standard 2Mbps connection you should be able to download one track every minute or a whole album in around ten minutes.

Tip: If you do sign up for eMusic, make sure you download its toolbar. It installs a toolbar inside Internet Explorer that allows you to search its music catalogue, and block pop-ups. Its best feature is a free music track you can download every day. As you'd expect some of the tracks are plain awful but some are fantastic and, just to remind you, they're free!

Audible

This small US-based operator has cornered the market in spoken word audio tracks and its recently launched a UK website (www.audible.co.uk). It's an excellent place to find spoken books that can help turn hours of boredom on a train or tube into an enlightening experience, with the right portable music player.

Service profile – Audible	
User	Beginner
What it is	An online service that specialises in audio books
Why bother	Like listening to spoken word books?
Source	www.audible.co.uk
Difficulty	Complicated
How long will it take to master	More than 20 minutes

To get access to this excellent service register as a user on the service. You'll then be asked to download one of the various software managers that run the service. If you have iTunes already installed you'll only need the basic Download Manager. If you're using Windows Media Player and want to download tracks onto a portable music player you'll probably need the Audible Manager. You can even download a program that will let you play spoken word audio tracks on a Pocket PC phone or Palm device.

Lets start with the Windows Media Player option of Audible Manager. Audible will now lead you through the various set up options – you can even let the installation process set up the software on your compatible portable music player. Eventually, Audible Manager will restart and ask you to log back in or set up a new account. It will also ask you to activate desktop player (in this case Media Player 11) and any portable MP3 device you may have set up.

Audible – AudibleManager

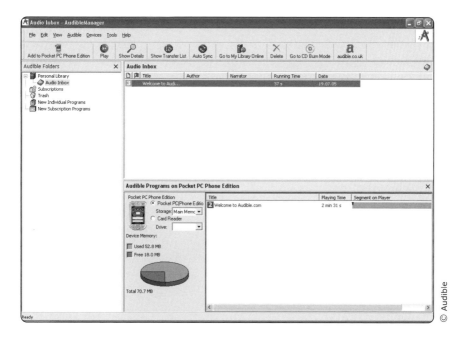

It's all a bit time-consuming and fiddly but eventually you'll have the program installed.

You next have to work out which subscription plan you're going for. At the time of writing you can either:

1. sign up to a basic package at £9.99 a month, which gets you a free book a month and one free magazine subscription; or
2. £14.99 a month which gets you two of each.

Once you've signed up for one of their subscription plans you need to log back into the main website and start searching for the audio books you want. When you've found one, buy the book and then apply any credits you are owed with your subscription. You go through the checkout process and eventually the spoken word book is added to your library.

Once you've made the purchase you'll be taken to 'My Account' where you'll see the book listed. Next to the book title you'll see 'Get It Now'. Click on this button and you'll be taken to a new screen where you can decide how you download the book.

Audible – My Library

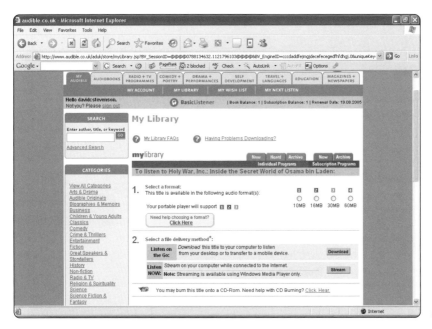

The book file is usually available in a number of different compression formats and file sizes. The best advice is to download the biggest, best quality file. You should also choose 'Listen' on the Go option.

A new downloading box opens up and eventually the book is loaded onto your computer – and onto a portable device if you have one that's been configured to run with Audible.

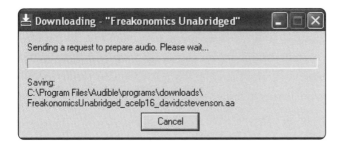

To listen to your book file on your computer, simply open up the main Audible program on your computer, select the book you want to listen to and then hit the 'Play' button on the top row. The audio file will now be played in your default media player (best to use either Windows Media Player or iTunes).

Audible – Audio Inbox

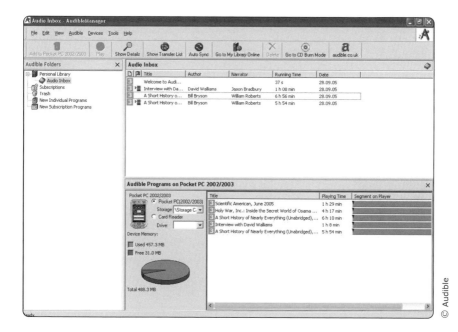

You can also burn the book files onto a CD if you want. Click on' Go to CD Burn Mode'. You'll now download a plug-in from Roxio that will let you burn your audio books onto a CD. But be warned, you'll have to restart your computer, and then activate the CD burner. It's a little time-consuming, but eventually you'll get there. Once you've restarted the computer and the program, click on 'Go to CD Burn Mode' and you should see the Roxio plug-in at the bottom of the screen. Now drag and drop any files you want to put onto the CD. Remember that you'll only have a maximum of about 1 hour 30 minutes to put on one CD. If your book is hours and hours long you'll need multiple CDs.

Other services

Wippit

Small, feisty music website, Wippit (www.wippit.co.uk) is by far the best of the also-rans of the digital music wars. Originally launched as a legitimate, paid-for rival to Napster-like file sharing networks it showed that you could build a network based on accessing other people's music files whilst staying strictly legal.

It's had a few ups and downs over the last few years and is still largely the preserve of indie-labels and bands, although some major music labels have signed up their music. But it still has one of the very best deals on the market. For £60 a year you can access its entire music library of over 60,000 tracks and what's more you can use these MP3 and WMA files in whatever device you want – none of this annoying shortlist of MP3 players authorised to use the service.

HMV Digital and Virgin Digital

Napster and iTunes are also facing more mass market competition from HMV Digital (www.hmvdigital.com) and Virgin Digital (www.virgindigital.co.uk). Both use their own players and both operate rather like Napster in that you can either pay per track or album or pay a monthly charge to access all the tracks as temporary downloads. As with Napster, this monthly subscription option is great value as long as you keep the subscription going – stop paying and the music tracks are disabled.

PlayLouder

PlayLouder (www.playlouder.co.uk) is the dark horse on the UK digital music scene. Until recently it was well known as the champion (like Wippit) of indie rock music, and many users simply joined the service so that they could access their Singles Club, which gave away free tracks every month or so. But PlayLouder has bigger ambitions and is launching a broadband ISP service that throws in access to all the music tracks in its library as many times as you want for as long as you want. Its clever trick is to ring fence all its users inside a kind of closed garden, a digital walled community, that PlayLouder can police. That means none of those naughty file sharers will be allowed on these networks and thus the big music labels feel safe in sharing their music for a fixed fee.

Which service to use?

The good news is that competition between these various online music services does seem to be encouraging innovation, although the price of albums and individual tracks does seem to be staying suspiciously stable and uniform across the various websites.

Range of music

The greatest progress has been made in opening up the record label libraries. You should now be able to find pretty much everything you'll ever need on one or other of the main sites. To test this out I set out to find ten, very varied tracks on three popular networks: iTunes, Napster and Wippit.

Song	iTunes	Napster	Wippet
Audio Bullys – *Shot Me Down*	✔	✔	✘
The Eagles – *Hotel California*	✘	✔	✘
America – *Horse With No Name*	✔	✔	✘
Ram Jam – *Black Betty*	✘	✔	✘
Bob Marley – *Pimpers Paradise*	✘	✔	✘
Dinah Washington – *Is you is or is you ain't my baby?*	✔	✔	✔
Dire Straits – *Brothers in Arms*	✔	✔	✔
Nina Simone – *Sinnerman*	✔	✔	✘
Harry Belafonte – *Scarlet Ribbon*	✔	✘	✔
Mr. Scruff – Ahoy There	✔	✔	✔

Napster seems to offer the greatest range of music, closely followed by iTunes with Wippit trailing well behind. To be fair to Apple's iTunes, it claims its library is slightly bigger than Napster's, but Napster does seem to have the edge on those 'most wanted' kind of tracks. Also Wippit's back library is much more indie rock focused and you wouldn't really expect it to have the same range as Napster or iTunes.

The number of no-shows at iTunes surprised me – not to have the classic Eagles' track *Hotel California* is a bit of an oversight, as is the absence of the Marley track. Still, the site seemed to have everything else you could possibly ask for and is obviously very strong when it comes to latest releases and specialist genres like jazz and spoken word books.

Cost

What about the cost of music online? Let's take Coldplay's popular album *X&Y*. On the high street the CD retails for around £10, but once you go online prices start to drop dramatically.

Shop/music service	Price inc delivery	Download service or shop
Napster	£6.95	Download
MSN Music	£6.99	Download
4CheapCDs	£6.99	Online Shop
101CD.com	£7.95	Online Shop
Tesco Jersey	£7.95	Online Shop
iTunes Music Store	£7.99	Download
Play.com	£7.99	Online Shop
CD-Wow	£8.75	Online Shop

Note: All the prices were correct at the time of writing.

The very cheapest way to buy this album is via a download service. Napster comes out top again, selling it for £6.95 as a download. MSN Music is not far behind at £6.99 but here's the really worrying thing: at least one online shop (not a download service) – low profile 4CheapCDs – is selling the album, the physical hard copy, for the same price as an electronic download (no CD, no CD case, no CD cover booklet)!

Ask yourself this: how can a digital music service charge the same as a shop that sells the physical thing? The shop-based retailers have to pay for, staff, warehousing, delivery and the actual CD and its duplication costs. It's also true that many of the cheapest shops don't have to pay VAT as they're based offshore, but that tax advantage should only account for a £1 difference in price.

Astonishingly, iTunes charge the same price for a music download as online shop Play.com. I say, astonishingly, because Play.com is a well known, well established, Jersey based operator with a great customer reputation, yet they charge the same price for this album even though they have to physically supply you with a hard copy.

And here comes the next outrage...

At least with a CD in your hands you can choose to do what you want with your music – if you want to copy it to your MP3 player, you can. Online music services still try to police your usage of the purchased album so that the tracks can only go on approved devices. Most sensible users choose to bypass this by burning the purchased tracks onto a CD, but guess who has to pay for the blank CD – you!

Digital music services do, to be fair, have a lot going for them. If you want to buy individual tracks (priced at between 79p and 99p) going online for a download makes absolute sense. Not all the tracks on Coldplay's *X&Y* are that great, so why not pay just a few pounds for the best tracks? Online music shops can't hope to compete with this flexible service.

The use of subscription-based services (pay £9.99 a month to access everything) is also a great deal, but be aware that doesn't mean that you're buying the music permanently. Stop paying the subscription and the music vanishes. Also remember that if you want to play these tracks on your authorised MP3 player you'll probably have to pay a premium.

Bottom line

Until the price of an album falls closer to £5, I'd buy the real thing, the CD, online or at a record store. If you want to see if you like the music before you buy the (hard copy, physical) album, sign up for one of the streaming services offered by the likes of MSN Music that lets you listen to each track on an album for 1p or through a monthly subscription plans that costs about £10 a month.

In this chapter we came across DRM, and where to download music and audio books. At this stage, you should now be feeling comfortable with downloading music from the internet and playing it on your PC. But strap yourself in, because next up is where the fun starts – video!

5

Digital Video

Words, sound and music are great, but they'll never measure up to the experience of video and film. The image accompanied by sound provides so much more satisfaction and information, yet it comes at a massive digital cost: size.

Video files are big

Video files are huge, so big in fact that until very recently working with video on a PC was solely the preserve of TV companies and video editors with giant super-powered computers with oodles of storage and expensive monitors. That has now changed, but there is one constant – regardless of its format, digital video gobbles up huge amounts of hard disk space and processor capacity.

Here's a small example.

Fetch one of your favourite pre-recorded DVDs and load it into your computer. Don't bother playing the DVD on your PC using software like Windows Media Player – instead go to the My Computer icon and locate the film DVD on the drive list. Right click on the Film/DVD icon and select 'Explore'. You'll see a list of folders. Look for one called VIDEO TS – this contains all the video and audio files. Open up this folder and then go to 'View' in the top right hand side of the screen and select 'Details'.

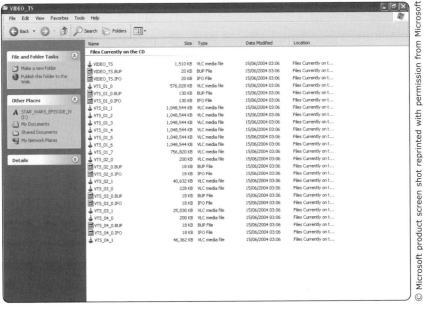

© Microsoft product screen shot reprinted with permission from Microsoft Corporation

You'll almost certainly see a list rather like the one on the previous page – from the *Star Wars IV* DVD. You'll see a series of files with odd names that, in this case, total 7.37GB. That's a huge amount of data.

To put it into some perspective, that works out at more than 15 times the data on a typical CD album (the uncompressed WAV files that is), or over 1000 times the total data from an album that been compressed into the MP3 format.

Or to put it another way – until just two years ago, that was only a few gigs under the size of the whole hard drive of a normal PC.

1 film = 1 computer hard drive!

This simple example underlines a number of important topics we'll be exploring in this chapter.

The biggest topic is that of compression – dump just a few DVDs onto your hard drive and you could run out of space in no time. That forces the computing industry to come up with ways of compressing the data.

But those different ways of compressing the video also present another challenge. There are many different compressions, codecs and systems for playing video, and many of them are so obscure and so tricky to use that the average user simply panics and runs away.

The message of this chapter is: *Don't Panic*!

Nearly all the major formats used in video are easy to manipulate and use provided you have a bit of know-how and the right tools.

But there's one last point. Even after using clever ways of compressing data, video files are still very large packets of data. That means they take a long time to stream or download and whenever you try to work with files, expect it all to take a long time. Hours, rather than minutes.

It's all worth it though. With just a small amount of knowledge and the right tools you too can turn your PC into a hugely powerful multimedia machine.

By the end of this chapter you'll have learnt how to:

1. Stream pop promos and short movie clips over the internet.
2. Work with clever codecs that turn movies into relatively small 600 Mb files.
3. Use simple to use media players that will play virtually every format.
4. Start backing up your DVDs onto your hard drive.
5. Get compressed movies onto a DVD that can then be played on your TV.

Video streaming

Go to the fabulous Atom films website (www.atomfilms.com) and witness a miracle. The idea of using the internet to watch video content was laughable until just a few years ago. Most of us were struggling along with old fashioned dial-up connections that made retrieving small emails difficult enough, without worrying about short video files.

Broadband has changed everything. Its not only made the process of downloading content much easier, its also prompted websites like Atom films (and the equally fabulous iFilm site) to innovate with new streaming technologies that allow reasonably large data files to be broken down into small packets and sent over the internet like traditional TV broadcasts.

We first ran into streaming with internet radio (check out the chapter on digital music) and the principle behind video streaming is no different. Just as music is chopped up and sent out in streaming packets of data, so is video. And just like streaming audio you need a special player on your computer to play that streamed content.

The first time you visit media-rich websites like iFilm and Atom, you'll be asked to specify which player you want to use. As you'll see from the iFilm window over the page:

iFilm – Media Preferences

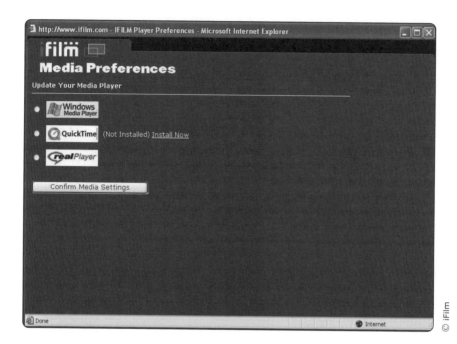

There are three main choices

1. **Windows Media Player** – much the best option
2. **Real Player** – the video-based client for the Real Networks
3. **QuickTime** – Apple's take on streaming video

I'd recommend that whenever you're given the choice use the Windows Media Player option. Having chosen which streaming player you use (you may also be asked to specify if you have a broadband connection or not) you'll then see a new Internet Explorer window pop up.

This in-built player contacts the main web siteservers and starts downloading video and audio content so that it can be played on your media player. The process can take anything from ten seconds to one minute and is called *buffering*. Eventually, you'll be ready to watch the video, but first you might be forced to watch some hideous US car advert (that's the price of accessing free video content these days). If the content is originally shot on video, the quality of the picture on the small screen will probably be rather poor while animation will tend to look very clean if a bit jerky.

Atom Films

Unfortunately, some web sites are a little more demanding than Atom and iFilm. Some insist on you using just one media player – either QuickTime or Real.

> **Tip**: Don't install either of these programs unless you really need to – they're both big and buggy and tend to insist on taking over your desktop. Luckily there are alternatives...

Real Alternative

We looked at Real Alternative in our chapter on digital music. It's a small, free program which doesn't have annoying spyware or adverts and lets you use all Real media content. You can find the program at most shareware websites, but its home is at www.codecguide.com/about_real.htm. Follow the usual installation instructions – making sure that the program associates itself with all Real files. Once installed it will play all Real Video files, no hassle.

QuickTime Alternative

Apple's QuickTime was, like Real Player, very much an internet streaming pioneer, but its been around long enough for someone to come up with a freeware alternative, called, unsurprisingly, QuickTime Alternative. It's available from the same website as the Real Alternative, at www.codecguide.com/about_qt.htm, although it's a much bigger file, weighing in at over 10MB. As with the Real Alternative, make sure the program associates itself with all QuickTime files (they all end in .mov) although this should be done as standard during installation.

Websites with great film content

- iFilm (www.ifilm.com)

  Fantastic website with loads of free videos, including mad spoof movies, real-life bizarre moments and some wonderful short clips.

- Atom Films (www.atomfilms.com)

  The best short film movie site on the net. Like iFilm you'll have to sit through the odd, really boring ad, but it's full of specially commissioned short films and animations of which the highlight has to be the truly wonderful Angry Kid.

- BBC (www.bbb.co.uk)

  What can I say – there's loads of video and programmes available here plus radio broadcasts.

- MyMovies.net (www.mymovies.net)

  Film clip preview service with all the best trailers.

Playing video files on your PC

You'll have worked out by now that Windows Media Player is a fairly useful piece of software that comes as standard on PCs equipped with the XP operating system. It plays most of the key video and audio formats and is, in the big scheme of things, fairly easy to use – it's also pretty much indispensable when it comes to video streaming.

VideoLAN Media Player

But Media Player is far from perfect, and although useful it shouldn't be your only multimedia video player. Why not try out its open source, free rival called VideoLAN Media Player (abbreviated to VLC Player) – you can get it from www.videolan.org/vlc.

Program profile – VideoLAN Media Player	
User	Beginner
What it is	A free media player that will play virtually every video format in existence
Why bother	Its very easy to use, doesn't use up too many system resources and is completely open source and reliable
Source	www.videolan.org/vlc
Difficulty	Very easy
How long will it take to master	A few minutes

It's hardly the catchiest of names but the VideoLAN player has a number of huge advantages over the Microsoft player:

1. It plays virtually every video (and audio) file format and compression codec on planet Earth with the exception of Real media files.
2. Unlike the Windows Media Player it will play all DVDs and SVCDs. The Windows Media Player can handle playing DVDs but it requires an extra plug-in, which has to be separately downloaded.

3. It's easy to use, has a very simple intuitive interface and doesn't come with any spyware, any annoying adverts to buy other Microsoft products or visit Microsoft websites, and doesn't come bundled with loads of annoying copyright protection tools.

After installation, start up the program and sit back and admire its elegant yet tiny interface.

The VLC player is a model of clever, restrained design.

To open a video file, simply select 'File', and then 'Quick Open File', find the piece of video and sit back and enjoy.

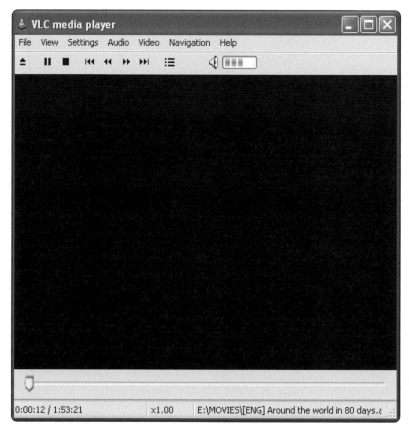

5 – Digital Video

If you want to play a DVD simply select 'File', then 'Open Disc'. Make sure that on the Disc Tab, the device letter is your DVD player and then press 'OK'. One word of warning though – some DVDs have a form of software protection built into them that stops you playing DVDs unless you install a specific piece of software devised by InterVideo. You can choose to install this software – it loads automatically when you insert the DVD – but my advice is not to bother as it's yet another piece of annoying and frankly unnecessary DRM, or copyright protection. The way around this program is to load in another very clever piece of paid-for software called AnyDVD – available from www.slysoft.com for $39. It stops all copyright protection software dead in its tracks and will let you play all kinds of DVDs.

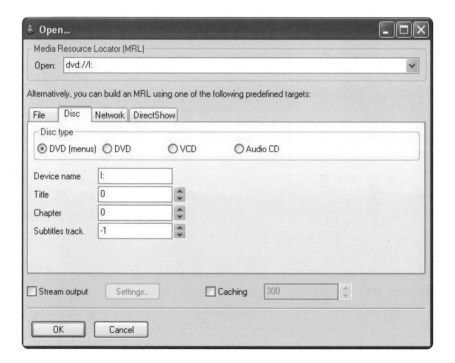

After you've loaded your DVD you'll be taken straight to the main menu page, where you'll see all the main options you'd get on a normal DVD player.

You probably won't want to watch the movie in a small box – why not watch it full screen on your monitor? To do this right click on the movie and select 'Fullscreen Mode'. You'll also see a number of other controls that let you alter the audio levels, choose subtitles and move on to the next chapter of the film.

If you want to get rid of the fullscreen, simply right click again and then unclick 'Fullscreen'.

VLC Media player

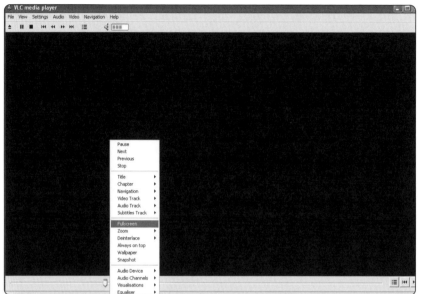

VLC Media Player has one last great strength – it'll play most of the key formats used in video compression including the dominant video codec which is called DivX (we'll take a closer look at this format next).

> **Tip**: VLC Media Player will also play any subtitles you may have with the film.

The Codec War - DivX vs XviD

One of the most annoying aspects of digital video is the astonishing number of formats and codecs in existence. Even the most experienced users are sometimes fazed by all the formats lurking around. And although players like the VLC Media Player will recognise most of them, there are some slightly weird formats (you might run into one called Matroska for instance) that even it will struggle with.

MPEG-1

The great grand-daddy of all video formats was something called MPEG-1 (the MPEG stands for Motion Picture Experts Group). This video compression standard was originally brought in so that the movie studios could fit a standard length feature film on one or two CDs, usually marketed as VCDs in places like Asia and Africa. The quality is close to VHS and, like VHS, it's a dying format with little or no content available these days.

MPEG-2

MPEG-2 is a distinct improvement on its predecessor and is the standard compression format used by most broadcasters and DVD manufacturers. If you use DVDs a lot, you're using MPEG-2.

MPEG-4

Demand for even smaller video files, with even more compressed data has spawned yet another mutation of this format - MPEG-4. This is an 'all things to all people' compression mechanism that is expected to replace MPEG-2 for broadcast quality applications, especially as it's approximately twice as efficient at crushing down file sizes. It has also been designed from the ground up to deal with streaming and interactive applications.

MPEG-4 comes in many different shapes and sizes, and one of the most common is something called DivX. This is a proprietorial video codec technology that's also based on the MPEG-4 standard. Its huge success led the developers to design a completely new product that's absolutely superb at compressing video to an extremely small size with a high degree of quality. The

DivX format has also gone through a number of different mutations – the first was something called DivX 3.11, which was then followed by DivX5 and now DivX6.

XviD

Just to really confuse things you might also see an alternative format called XviD. It's not actually a commercial product like DivX, but an open source project which is developed and maintained by programmers all around the world. XviD, is yet another take on the MPEG-4 format and like DivX is expert at compressing big film files ripped from a DVD into small files.

AVI

You'll also see yet another file type lurking around called AVI. This is also not actually a video compression format as such, just a file type and stands for Audio Video Interleaved. Its Microsoft's own video format combining audio and video coding in alternate segments.

VOB and IFO

Confusingly if you explore the files contained within a DVD you won't see any mention at all of the above AVI or MPEG files – instead you'll see lots of VOBs and IFOs. These aren't actually compression formats (like MPEG) but simply file types that marshal together all the video, sound and subtitle files in one folder that can be played on a DVD player – the VOB files (VOB stands for Video Object) contain the raw video and audio data, while the IFO file is just an InFOrmation file that tells the DVD player where to find video and audio files.

WMV and MOV

In the world of streaming video you'll also come across WMV files (the Windows Video streaming standard) and MOV files, Apple's take on video streaming files using the QuickTime format.

Format	Amount of MBs storage needed for 60 minutes of film	Can you play on an ordinary DVD player?	Video quality	Dolby Digital Audio
DivX 6	660	Yes but only a small number of specialised devices	Pretty good	Yes
XviD	660	Yes but only a small number of specialised devices	Pretty good	Yes
MPEG-1 or VCD	600	No	Poor	Yes
MPEG-2 or VOB files	2,600	Yes	Superb	Yes
QuickTime	750	No	Poor	No
Windows Media files	880	No	Poor	No

Source: Toms Hardware Guide – www.tomshardware.com

Compressing video – the DivX/XviD revolution

In the Old World of home entertainment, pre-broadband, we were, by and large, satisfied with watching films on a DVD, on a DVD player, in our lounge. Broadband has changed all that. More and more of us now prefer to either play our films on a PC, and perhaps source our films over the internet via some form of download network.

DivX is the software-based backbone of this digital film revolution. It's essential you understand it and use it for two very simple reasons:

1. If you want to download movies from the internet file sharing networks we'll encounter in the next chapter, you'll need DivX. Most of the film content online is compressed and encoded using DivX or its open source rival XviD. If you have the right codec you can then play these compressed

movies back on either your PC or on a DVD player that has built-in DivX/XviD support.

2. If you want to store your favourite movies on your computer you'll need to compress them. Most DVD-based movies contain huge amounts of data – between 7GB and 9GB – which can only be stripped down and compressed using a codec like DivX. Once they've been squeezed down to just under 700MB you can then pack dozens of movies onto a dedicated hard drive that stores all your films.

Working with DivX and XviD

Most of us only ever require one thing when dealing with either DivX or XviD – the codec. This is the small bit of software that tells your media player how to deal with all this compressed video.

The good news here is that the VLC Player already has the codec built into it so you don't need to download any new software. Microsoft's Media Player 11 Series doesn't come with built-in DivX and XviD support, but as soon as you try and play the files the program will head online and download the appropriate codec automatically.

If any of your media players still can't play DivX and XviD files it's easy to find them online and download for free, although you should be able to get by for the time being with DivX 5 codec (it comes with a player). This is available for free at www.divx-digest.com/software/divxcodec5.html. Just download the Free Standard version. If you want the most up-to-date version of the DivX software go to the company's website at www.divx.com/divx/play/ and download the free Play bundle. The XviD codec can be found at www.xvidmovies.com/codec/. Simply select the latest Koepi code.

That's it: any DivX or XviD file should automatically play on your computer, hassle free.

DivX Create Bundle

You can choose to use DivX simply as a codec that works with other media players, or with DivX's own player. But there is a far more powerful tool lurking inside its website. It's called the DivX Create Bundle, and although it's not free (it costs $20) it gives novice users the chance to create their own DivX files. It also contains the latest version of the DivX codec – version 6. According to the developers this new codec offers up to 40% better quality and compression than the (previous) DivX5 codec, plus the ability to work interactive video menus, subtitles and multiple audio tracks. It's all useful stuff but the most powerful tool is the DivX Converter. This is a great way of taking movies off your DVDs, and then compressing them into smallish data files – between 600MB and 800MB – that can then be stored on your hard drive, or even burnt onto a DVD and then played on specially adapted DivX DVD players.

Program profile – DivX	
User	Beginner
What it is	A program that converts your video into compressed files that can then be played on a PC
Why bother	Compress a film down to 700MB and you can back-up dozens of movies
Source	www.divx.com
Difficulty	Very easy
How long will it take to master	A few minutes

DVD Decrypter

If you do decide to use the software I'd also recommend downloading an additional program called DVD Decrypter – available for free at www.dvddecrypter.com. This clever piece of software grabs video files from a DVD, breaks through any encryption that might have been built into the DVD and then dumps the video files onto your hard disk. Why the need for extra software? The DivX Create Bundle and specifically its Converter program are not designed to work with copyright protected DVDs that have special software built in to stop any form of backing up or copying – that means that if you load in a copyright protected file they won't be able to do any compression, which is where DVD Decrypter comes in..

Install the program, and then insert the DVD you want to back-up into your main drive, and then launch Decrypter. Before you start you need to make sure it's going to decrypt and extract the right files – click on 'Mode' and make sure you have ISO selected.

DVD Decrypter

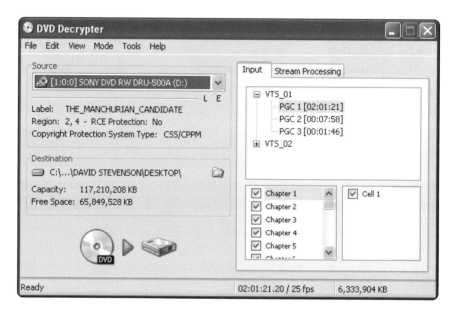

You'll see that it's automatically selected all the main film files that contain the movie.

All you need do next is select a folder to save the decrypted files to and then hit the main 'DVD to Disk' button on the bottom left hand side. The program will now take about 30 minutes to extract and unlock the files.

DivX Converter

Once you've finished it's time to open up DivX Converter. As software goes this is a rather pleasing minimal interface designed rather like a stopwatch. Select the Home Theatre profile by using the arrows at the bottom of the screen. Then press the 'View List' button. A new 'Files to Encode' box appears. Simply navigate your way to where you've saved the files that DVD Decrypter extracted and cracked by using the 'Add' button, then click OK. You'll also be asked on the View List screen if you want to convert the files to a certain size – 700MB for a CD, or 2GB for a DVD. Choose the 700MB/CD option.

DivX Converter

DivX Converter will now take these video files and compress them – the amount of encoding time will vary but it shouldn't take more than three to four hours.

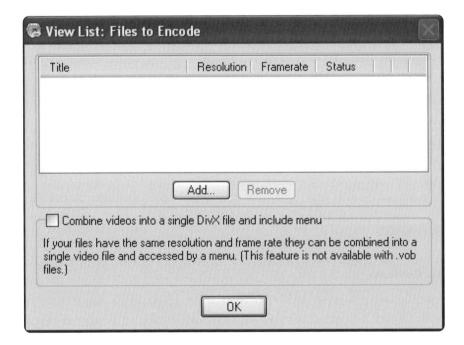

> **Tip**: Simple, elegant interfaces such as DivX Converter are great, but they have one big downside, namely you don't get many extra features, especially the ability to set up any individual options and preferences. One of the slightly annoying features of the Converter tool is that it doesn't tell you where you're saving your files to – it just starts converting. It actually saves them in the DivX Films folder – you'll find it inside the program files (click on 'My Computer', pick the main drive which is usually C), look for DivX and you'll see another folder called Movies. This is where your compressed film is saved to.

More weird file types

The name of the game with video, as you'll have guessed by now, is to find ways of compressing the size of the files so that they can be easily transferred onto a computer and then transferred across the internet. For most people who use file sharing networks, like BitTorrent and eMule, that means working with DivX and XviD. But you'll be slightly disheartened to know that there are in fact yet more formats and file types that are widely used. Two in particular stand out: RAR files (a kind of exotic zip file), and Bin/Cue files.

RAR files

Programmers have been trying to find clever ways of compressing files for decades. Unlike modern compression technologies, such as DivX, zip software doesn't alter or convert the underlying video files. It simply uses clever algorithms to compact down the actual file itself. Zipping is very easy, and support for unzipping zipped files is built into the latest versions of XP – you simply right click on the zipped file and select 'Extract'.

This simplicity has been taken to rather extreme lengths with a format called RAR. You'll find all sorts of RAR files on the internet, but most of them will be video files – it's a great way of zipping up big film files and then turning them into lots of little data packages that can be broken up and sent over the internet. WinRAR is a shareware program (you get a 30 day free trial) that will handle formats such as ACE, ARJ, BZ2, CAB, GZ, ISO, JAR, LZH, TAR, UUE and Z.

Go to www.rarlab.com and click on 'Downloads'. You'll now see a list of all the available versions – click on the latest version.

WinRAR – Setup

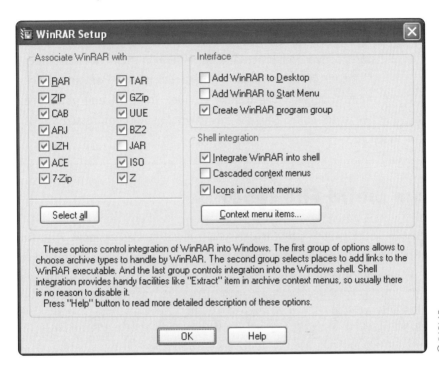

During the installation you'll be presented with a box that asks whether you want WinRAR to manage a huge range of archived, zipped and compressed files. Keep all of them ticked.

Open the newly installed program and you'll see the main screen.

WinRAR – Main screen

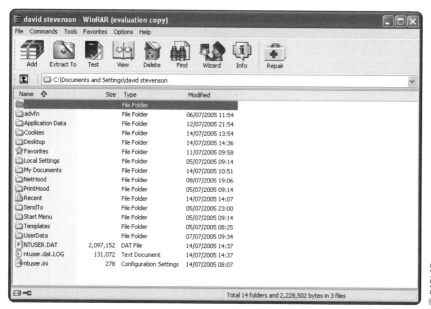

If you download a file with a lot of .RAR files, click on the top folder in the folder list. Alternatively use the drop-down menu line – in the diagram above the one starting C:\Documents and Settings\[Your Name], and pull the list of locations down.

Once you've navigated your way to the .RAR file, highlight it and then click the 'Extract To' button at the top.

WinRAR – File extraction

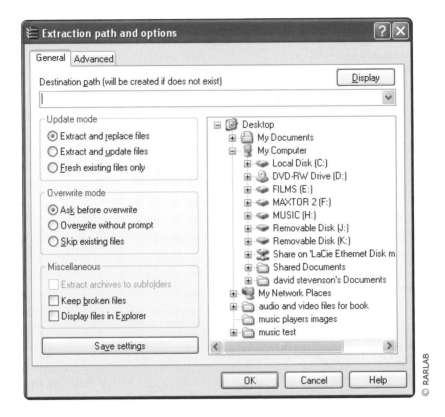

This Extraction Box will give you a number of options for extracting the .RAR files. First set the location of where you want the final file saved. Then leave all the other options as is and hit 'OK'.

WinRAR will now take a minute or two to extract all the files into your destination folder.

That's it! Not at all complicated.

Cue and Bin files

You may encounter a strange pair of files in your online travels, namely .cue and .bin files. They're brothers and sisters of a rather exotic little format that originated with CDs. Many programmers found that the best way of dealing with a programme or film was to turn it into an image file that could then be burnt as one onto a blank CD. If you come across these files, you've got two choices:

1. You can burn them onto a (V)CD which can be played back on a home DVD player. To do this: open up a CD burning program like Nero (or any other CD burning software that can burn from image like Alcohol 120%) and select the 'Burn from image' option. Select the cue file you want to burn, and burn it to a CD or as (S)VCD. Afterwards, test the CD and make sure it works; if it does you can delete the cue and bin files from your PC.

2. You can convert the image file into a number of ordinary program files by opening it using special software like VCDGear (it's free), which then converts it into a standard movie MPEG file (it will even convert into DivX for you). To download the program go to www.vcdgear.com/download.html and download the latest available version – v3.56 last time I looked. When you have successfully converted the cue and bin file to .mpg, you can delete the cue and bin file if you don't intent to burn them to CD. (You can also burn the .mpg files as (S)VCD with Nero.)

VCDGear – Main page

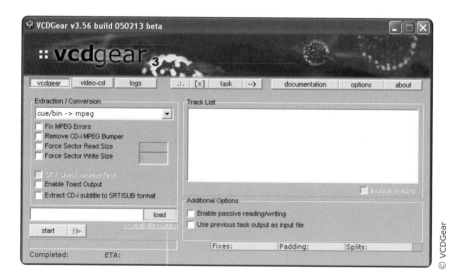

VCDGear

VCDGear is incredibly easy to use. Just click on the 'load' button to find the cue and bin files. Then tell the program that you want a cue/bin to mpeg conversion (the standard option), then press 'Start' and eventually your cue files will be converted into a nice, easy to use MPEG file that can be played as normal on any computer.

Daemon Tools

If it's not a movie but a game that's been downloaded and compressed you can download an alternative free tool called Daemon Tools: www.daemon-tools.cc/

Open Daemon Tools, click the icon in your system tray and select 'mount image', then select the .cue file and it will be mounted to a virtual drive (this means your PC thinks that it's an actual CD in the CD tray, but in fact it's a .cue/.bin file on your hard drive, opened by Daemon Tools and mounted as a virtual drive).

Playing video files in the living room

All this talk of codecs, weird file types and compression, slightly obscures one nagging problem: they're all PC-based. How do you watch all this compressed content in the lounge, on a TV, rather than on a PC?

Like many, I do have a PC with a good LCD screen, but I much prefer relaxing in the lounge (or on the hammock) a long way from the study where the PC lives. Surely, the big risk is that all these clever wizzy codecs and compressed film files will be doomed to sit chugging away on our PCs, never to emerge in the lounge living area in a viewable format?

Step forward the next generation of smart home electronics devices. I've selected three rather clever devices to show what's ultimately possible with a small amount of money and common sense. With each and every one of these devices you can release your compressed DivX files from their PC home and start watching films in your living room.

I've detailed each of the products, with a rough guideline price from my favourite cheap electronics website (www.ebuyer.com), but don't get too hung up on the actual specific devices (or the website as there are plenty of good alternatives). The point is that each of these *types* of devices can play compressed video in your living room.

LiteOn DVD player

The LiteOn DVD player is perhaps the easiest way of getting DivX (or XviD) material into your living room. On one level it's a simple DVD player – at the back of the player, there's a scart lead that connects into your TV. All you need do is put a DVD in, switch on and it'll play.

The trick, of course, is to get the DivX files onto your DVD. To do this, simply find the DivX files on your hard drive, then transfer the files into a burning program like Nero. To do this, open up Nero Burning ROM and then select a New Compilation. Make sure it's set to DVD ROM (UDK/ISO), then drag and drop the files onto the blank disc. Next, select the 'burn' icon. You should be able to fit 6 DivX films on one blank DVD disc.

LiteOn LVD-2010 DivX DVD Network Player Inc Wi-Fi Adaptor.

Price on www.ebuyer.com at the time of writing: £85

This LiteOn player also hints at another possibility, using the WiFi adaptor to stream files over your own home network. I wouldn't recommend this option for novice users but the principle is the same – instead of dumping your DivX videos onto a disc, a server that works inside the machine goes through your home network and finds the DivX files. It then streams over the network onto your TV.

> **Tip**: Make sure that your DVD player can play both DivX and XviD files. This should come as standard in this sub-market of smart DVD players but make sure before you buy.

LaCie Silverscreen

The one big downside with these cheap DVD players (some retail for as little as £30) is that you still have to use blank DVDs. You can of course use recordable DVDs, burn the files on, and then erase them after you've watched them. But there's still this problem of fiddly discs lurking around. One way around this is to use mobile hard drives that connect to your TV in exactly the same way as a DVD player. The LaCie drive is simply the best of the bunch but also the most expensive – you can buy these systems for under £100.

These are incredibly simple to use. It's a hard drive that has a USB slot in one end and a slot to connect to a TV in the other. You load it with DivX and XviD films from your PC (an 80GB drive should take over 100) and then disconnect it, walk into your lounge, fix it up to the TV using audio visual leads (scarts) and then switch it on and navigate through the menu to find your film.

Sounds too good to be true?

It isn't. These devices work brilliantly although, in my experience they can be a little iffy playing some files that haven't been correctly encoded.

Creative Portable Media-Centre 20GB MP3 Mpeg

Price on www.ebuyer.com at the time of writing: £185

It's a fairly similar story with the last device, one of the new generation of portable media devices that operate in the same way but are designed to be used out and about. They also feature a hard drive that connects to a PC for uploading films, and they also have a way of playing the content on a TV. The clever bit though is that they have an in built battery power supply, which means you can watch the films out and about using the very nice colour screen. It's the most expensive option but it's truly portable and you can play music and look at photos. The only downside is that not all these portable media players can handle DivX movies.

Backing-up your DVDs

Would you like to back-up your DVDs?

Forget about it unless you have the right tools.

Luckily, these tools are often free, and easy to use, although you won't find them heavily advertised outside techie circles. Applications like DVD Shrink or ratDVD suddenly open up a whole new multimedia world – they give you the opportunity to back-up all your DVDs onto a single external hard drive that can be used as the basis of a new home entertainment centre of the future.

What do you need?

Not very much, except a bit of clever software and a DVD burner. The DVD burner is an obvious accessory, plus one of three different programs. In this chapter we'll talk you through each of these tools, all of which have a slightly different approach to backing up your films.

- DVD Shrink

  The wonderful DVD Shrink will compress your movie down to a file size that can be burnt onto a back-up DVD or filed on a hard drive.

- Gordian Knot

  This tricky but fantastic little program will compress your DVD into a single compressed file using DivX that can then easily be stored on a hard drive.

- ratDVD

  The even trickier, but even more spectacular, ratDVD will do the unthinkable. It will take all the extras and menus on a DVD plus the film, compress it to heck, and turn it into a 1GB file that can be stored by the dozen on your hard drive and used for later playback.

The above three programs are free and should take no more than an hour or so to master.

Some DVD basics

You need to understand two simple but not insurmountable problems.

- Your movie DVD cannot be copied straight onto another DVD using a burner

  Stick the film DVD into your hard drive and go to 'My Computer' and right click on the icon that shows the movie DVD. Select 'Explore' and then check how many gigabytes this movie is using on the disk. You'll be surprised to learn that even though DVDs have a stated capacity of just under 4.5GB, your movie is in fact well over 5GBs. Clone or copy it onto another disc and it just won't fit. That's because film companies use special technology that means that most DVDs actually store between 7GB and 9GB. Luckily all the programs in this chapter use powerful compression technologies that squeeze the original files down so that they total less than 4.5GB. That means the compressed files can then be burnt onto a blank disc.

- Most DVD films are encrypted

  Your favourite film DVDs will probably boast some form of copyright protection, which prevents you from either copying them to your hard drive, or, if you do manage to do so, stop them from playing the resulting files. This protection changes over time and the names used by the encryption engines vary considerably (you might see Macrovision mentioned a lot), but the good news for most consumers is that they've all been cracked at one time or another. Two programs stand out as brilliant decrypters – AnyDVD and DVD Decrypter.

Step 1 – Cracking DVD encryption

Want to learn how to crack the copyright protection on DVDs?

AnyDVD

Your first step is to visit the site of software vendor SlySoft (www.slysoft.com) and browse through the features of AnyDVD.

It's the only software in this chapter that's not free – it costs just under $40 – but it's worth every penny. Let's call it the best bit of stealth software you'll ever need, and best of all it's available on a free trial for 21 days.

Install the program as normal and you'll be surprised to see that AnyDVD never really does anything obvious on screen. That's because it's a background program that systematically strips all DVD files (and audio music CDs for that matter) of their copyright protection on your computer. Shove in a DVD film and it will play and copy across perfectly with no annoying copyright warnings.

DVD Decrypter

DVD Decrypter is the freeware rival of AnyDVD. Its great advantage is that it is free and it's astonishingly easy to use. But its developers aren't upgrading the software and it's only a matter of time before it stops working as new film titles boast new layers of copyright protection. At the moment though it's a fantastic bit of technology that works perfectly.

Program profile – DVD Decrypter	
User	Experienced
What it is	A free program that rips copyright protected movies onto your hard drive
Why bother	It bypasses all the DRM that stops you copying movies for back-up purposes
Source	www.afterdawn.com
Difficulty	Easy
How long will it take to master	Ten minutes

What does it do?

Unlike AnyDVD, it doesn't operate quietly in the background, running unobtrusively as an adjunct to media player programs. It's designed to rip the film files of a DVD, and dump them into a folder on your PC, whilst also breaking through the copyright protection on the DVD. To download it, visit the www.afterdawn.com website and search for the latest version of the software.

- Once installed, place a DVD in your main PC drive and open up DVD Decrypter.
- After a few seconds reading the DVD, the main DVD Decrypter screen pops up.
- Click on the 'Mode' menu at the top of the screen. You'll now see three main choices – let's concentrate on the top two, File and IFO. If you choose File, DVD Decrypter will take all the files off the DVD and dump them in a specified folder. This is great if you want to copy all the extras and menus but the resulting files will be huge! The IFO option automatically targets the files on the DVD that contain only the main movie (which is usually broken down into a series of 1GB files).

5 – Digital Video

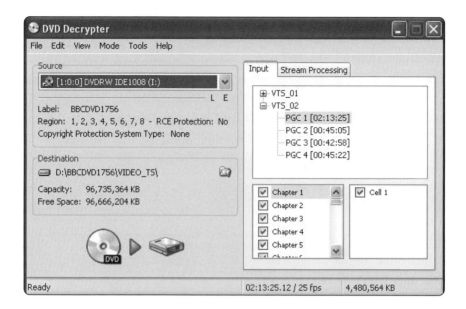

- When you've chosen which set of files to copy across, simply click on the little folder icon under 'Destination' and select where you want to save the files to, and then hit the 'DVD-to-Hard Drive' icon and DVD Decrypter goes to work – roughly 30 minutes later the file has been copied across and all copyright protection removed.

- But be aware that once you've ripped the files you still need another program to do something with them. Whichever Mode you choose you'll still have files that are too big for any blank DVD to handle. You'll need one of the next bits of software to convert the files into something that's usable.

Step 2 – Compress the files to burn onto a blank disc

DVD Shrink

DVD Shrink is quite the easiest piece of video editing software you'll ever run into. It uses DVD Decrypter to rip the content of the DVD movie and then special software to compress your film files into a folder that's small enough to fit on one single blank DVD.

Program profile – DVD Shrink	
User	Experienced
What it is	A free program that rips copyright protected movies onto your hard drive and then shrinks them into a file that's big enough to fit on one blank DVD
Why bother	It bypasses all the DRM that stops you copying movies for back-up purposes *and* then compresses the film files
Source	www.dvdshrink.org
Difficulty	Easy
How long will it take to master	Ten minutes

To get the software, simply visit its main home page at www.dvdshrink.org. Click on the US/UK flag and it'll take you to the home page of this rather wonderful little tool. This eminently sensible little website doesn't bother with huge long tomes about its software – it knows it's great, easy to use software, so simply follow the links to the download page.

- Click on the 'Where' button and you'll see a Google search box to help you find a site to download it from. It's worth noting that the volunteers behind DVD Shrink have stopped development work on it so 3.2.0.15 is probably the last reliable version of the software.
- Click on any of the mirror sites and, in a few minutes, you'll have the software ready to install.
- Install the software as usual and then insert a DVD you want to copy.
- Now run the software.
- Click 'Open Disk'. A new screen will appear – this says 'Open DVD' and gives you a short title for the DVD. Click 'OK'. The software now analyses the DVD. Tick the 'Enable Preview' box and you'll see it whizz through the various video components including the opening menu.

DVD Shrink – Analysing

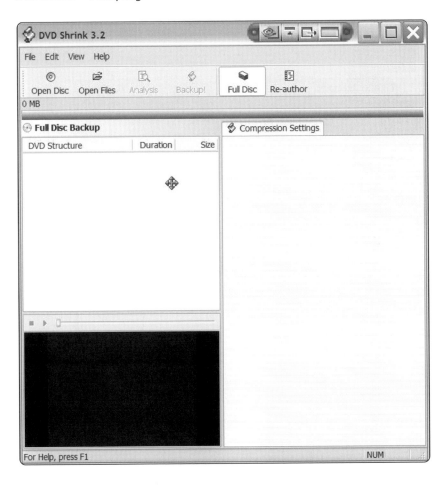

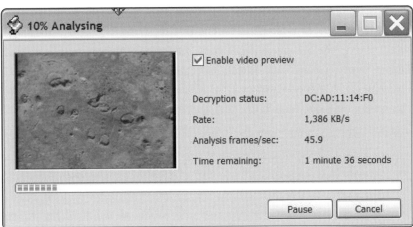

- Eventually, after about two to three minutes, it will finish analysing the DVD and our original screen with new details added to it will appear.

DVD Shrink – Details

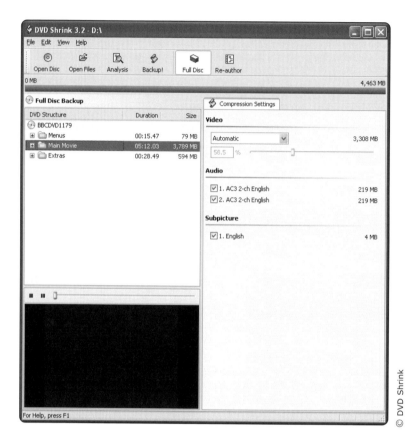

- You'll now be given a bunch of different options. The key bit of information is at the top – the big green line. Hopefully this line will be entirely green with no red – that means DVD Shrink can compress the files below the crucial 4475MB limit. This is the capacity of most consumer DVDs used for burning – anything above that and the content won't fit onto a standard DVD.

- On the right side you'll see a box with 'Compression Settings' on the top. All you care about is that it says 'Automatic'. Below that you'll see a percentage. That tells you the amount of compression in use – if it says 58% for example, that means that it's compressing to 58% of the size of the original and is well within the safe limits of compression. In my experience you're OK down to around 40%.

- Next you'll see various audio options. This is simply the list of all available audio streams. In some cases you can have the normal 2 channel plus further surround sound options. You might also have options for foreign languages – if you do, you can safely ignore these and uncheck them.
- Lastly, there's the sub-picture options, which are the subtitles. They don't take up much space so keep them.
- On the left side of the screen you'll see some little boxes that include 'Menus' (the various opening menus of the DVD), the main movie, and any extras like 'The Making of...' You can choose to have all or none of these by unchecking the main movie. If you click on the various boxes you'll see all the various episodes/options revealed. Click on any of the options you want.

DVD Shrink – Backup DVD

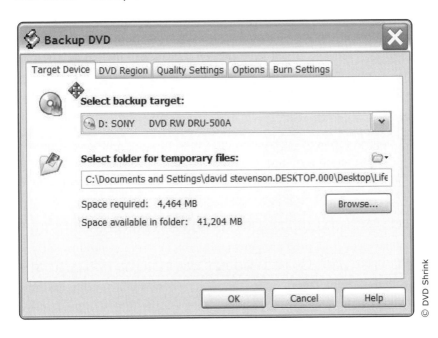

- When you're done click on the 'Backup' tab at the top. You're now presented with a new bunch of options. The first line tells you which DVD burner you'll be using – if you have Nero, DVD Shrink will manage the burning of the new back-up disc as well. In the absence of Nero, you can simply shrink the files into a folder and then use another burning application to handle the ripped files.

- Next you can set the 'temporary' location of the Backup files on your hard disk.

- The DVD Region option allows you to make your new back-up disc region free, thus avoiding any future headaches if you happen to take the DVD abroad for instance.

- In the Quality Settings area you'll be presented with two main options: Perform Deep Analysis and Error Compensation. The deep analysis option will encode your DVD in two passes. This will take significantly longer, but it may be worth it because it will certainly improve video quality by allowing DVD Shrink to better distribute the required compression throughout the various scenes in your DVD and it will ensure that DVD Shrink accurately meets the target size.

- Error compression is used to clean up the final picture. When video is compressed, small errors or artefacts are introduced. This is an unavoidable consequence of video compression, and DVD Shrink can't stop this from happening. However, it can keep these artefacts to a minimum. The best option is probably the default one, Sharp, which will tend to preserve the sharpness of the original video.

- If you are concerned about speed, don't enable any of these options – your back-up will be completed in just over an hour.

- The last few options are straightforward. In 'Options', make sure the program runs in Low Priority and in Burning check that it'll run at maximum speed.

- You are now ready to go! Press 'OK' and DVD Shrink does its thing. It will probably take at the very least a few hours. My advice is to run this overnight.

- Eventually you'll be prompted to supply a blank disc for burning – if you left the main program running overnight, you'll probably discover it's been looking for a blank disc for hours, but don't worry. Simply open up the disc drive, take out the original DVD and put in a blank. It'll then see the new disc and start burning.

Gordian Knot

DVD Shrink may be wonderful but its final output – the compressed files – will still be huge, many gigabytes in size. Also the final result will be a classic DVD – it plays like any other DVD. That's great if you want to play it in a DVD player, but it's less useful if you want to store a back-up of all your best DVDs on a hard drive.

Program profile – Gordian Knot	
User	Very experienced
What it is	A free program that will manage the whole process of converting your DVD films into a single 700MB DivX/XviD file
Why bother	It manages all the ripping, bypassing copyright protection and compressing
Source	gknot.doom9.org
Difficulty	Complicated
How long will it take to master	One hour plus

If your end game is to store dozens of movies on one 80GB hard drive (you could fit a hundred or more compressed films on a single back-up drive) your best bet is to use compression software that shrinks a multi-gigabyte movie into a 700MB file using either the DivX codecs or XviD. Step forward the complex but elegant Gordian Knot.

This free software started out as a simple bitrate calculator for DivX encoding, but has evolved to become an integrated package, or 'front end', for the entire process of DivX/XviD encoding. In fact, when you download the software, you're actually downloading a whole collection of packages and codecs that have all been integrated into one user interface.

Warning: As software goes this is quite complicated. It's not one of those desperately easy 'click this button' pieces of software. But it works and once you've mastered the process you'll find it awesomely powerful. Only for experienced users!

Downloading and installing

The first step is to download this free software.

Go to gordianknot.sourceforge.net and click your way through to the Downloads section. There you'll see the latest file releases in the centre of the page. The Rip Pack, includes a number of bits of software that will manage the process of copying across the DVD files, decrypting them and then converting them into a video format that can then be compressed. The second package, a codec pack, contains all the tools you'll ever require to compress the files. Download both.

Install the Rip Pack first. On set up you'll come to a screen that prompts you to choose the different components of the installation. Keep them all ticked and press 'OK'. The set up software will now install the various components that include mysterious items like BeSweet and VobSub. Don't worry too much about what's being installed, it's all incredibly useful!

Next, install the codec pack that controls the compression side of the package. Again, a number of packages will be installed – click yes to them all.

After a few minutes all the packages will be installed. You're ready to compress your films!

First time set up

The first thing to do is to make sure that all the programs that work with Gordian Knot are properly installed and that Gordian Knot knows where they're located.

Start Gordian Knot up. Find the program paths tab at the top of the screen and make sure that all the key programs have a proper pathway specified – if they're not there you can simply use the 'Locate' button to go into your programs folder on your hard drive and locate the .exe file that launches the program (usually it has a small symbol telling you what this .exe file is).

Now that you've checked all the programs are properly integrated into Gordian Knot, let's click back on the first tab on the top: Ripping. You now have two choices – you can use 'Rip & Process' to process the files on the DVD in 'one pass'. Effectively, this little tool runs the whole process for you. Personally, I don't use this function as I found it just as easy to run all the key ripping programs individually to make sure that everything is working fine.

Compressing in four easy steps

Step 1 - Rip files from the DVD

The first challenge is to work out how to rip files from the DVD. We've got to get the files off the DVD and onto our hard drive and also get around the encryption so that we can store video files on our computer for later processing.

Gordian Knot - Ripping page

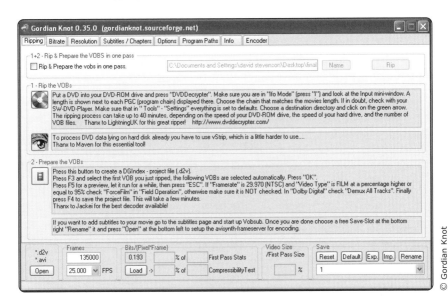

Click on the gold disc under 'Rip VOBs' (the technical name for the format DVD films are stored in). DVD Decrypter now starts up.

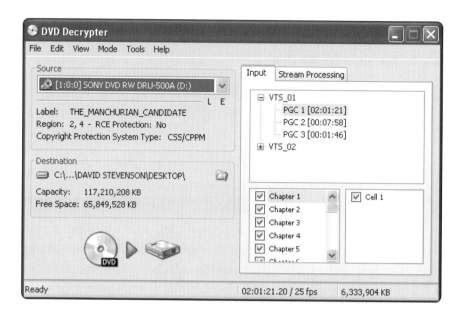

The program reads your DVD and works out what's on the disk – on the right side of the screen you'll see the various components under VTS. The software will usually choose the longest bit – the film – plus, in a box below, you'll see all the chapters ticked (they should all be ticked, just make sure). Problems can arise if there are, say, two movies on the same disc or if there's an 'extra' – say a film about the making of – which is as long as the actual film. If this is the case run the DVD and make sure you work out what exactly is the main film file.

Next, you choose the location of the ripped files on your computer. I usually create a folder on the desktop with the name of the film – if you want to do this just double click on the little folder icon and tell the program what folder to save it in.

Now press the big button at the bottom on the left side with a DVD symbol, an arrow and the symbol of a disk drive. The program will now take something like 30 minutes to rip and decrypt the film. After it's finished just check in your chosen folder to see if the VOB files are all listed in there.

Step 2 – Turn it into a vStrip file

Next go back to the main Gordian Knot window and click on the icon below the DVD Decrypter – the symbol with an eye called vStrip. This little program gathers together all the files you've ripped and puts them in the right order, in one, single file.

Click on the main vStrip button. You now need to add the files you've just ripped onto your hard drive. Locate them using the 'Add' tab midway down the page and pick all the files you've just ripped.

Now look at the top of the page again and you'll see a second tab next to the Input tab called IFO. DVD Decrypter also ripped a special little file called an IFO file, which is a bit like a music playlist. It's a file that tells the DVD machine in what order to play all the components of the DVD – the main film, the extras and the chapters within each of these. You need this so that all the VOB bits on your hard drive play in the right order. Go to where you saved the VOB files and you'll see an IFO file. Add it to vStrip.

Lastly, go back to the top of the page and locate the Output tab (next to the IFO tab). Click on it and you'll see where vStrip is about to save the files to – Output Name. Use the button here to give it a name. I usually call it the vStrip file and put it in the same folder as the rest of the ripped files. Make sure the Remove Macrovision box is also ticked (this removes the copyright protection) and press Run. In about 20 minutes your files will be ready.

> **Tip**: Because DVDs have so much content on them beware that you are ripping the main feature. It's not always the case that the VOB files marked VTS_01 are the right files – the film company may have called the main files VTS_03, or frankly anything under the sun.

vStrip – Main page

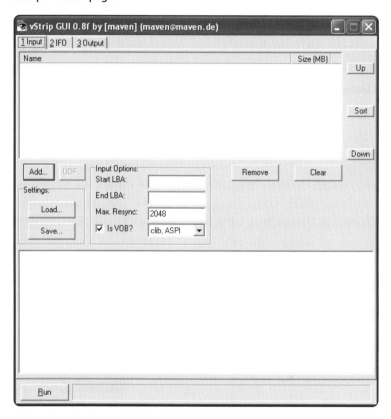

Step 3 – Prepare the vStrip file for compression

You should now have a big vStrip file in the main folder you've ripped your film to. The next stage is go back to the main Gordian Knot front page (to our ripping section) and go down the page to the 'Prepare the VOBs' section which boasts a small button with a brightly coloured strip of film on it.

This program now gathers together the vStrip file and other bits and pieces and puts them together in a special file (called D2V) that can be used by compression software. The instructions for the use of this quick little program are on the main screen but I'll just recap them.

- Press 'F3' and select the vStrip file you've just created. Click 'OK'. Now press 'F5' for a preview, let it run for a while to make sure it looks okay, then press 'ESC'.

- If Framerate is 25 FPS (PAL) and Video Type is FILM at a percentage higher or equal to 95%, check 'ForceFilm' in Field Operation, otherwise make sure it is *not* checked. In Dolby Digital check 'Demux All Tracks'. Finally press F4 to save the project file. This will take a few minutes.

Step four - Compressing the files

OK, so we have our D2V files saved in a folder. It's now time to compress these huge files down to 700MB or even less.

At the bottom left hand side of the main Gordian Knot window you'll see a button marked Open. Select the D2V file you've just worked on. A preview video screen will now appear with an image from around the middle of the film – minimise this for now. Now go back to the top of the screen and look for a tab marked Bitrate. A new screen will appear.

Gordian Knot - Bitrate page

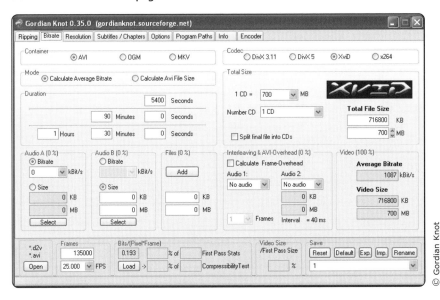

There's loads of information again here but the relevant bits relate to the codec you're going to use. It's basically a question of working out which codec you want: XviD and DivX 3.11 or 5? Most specialist DivX compatible DVD systems will play all types of files. In my experience the safest bet is to use DivX 3.11 (the most compatible with different players), the best looking video is DivX 5 and XviD is a great all round alternative. Whichever format you choose, you can effectively leave all the other boxes as they are.

225

Using all the presets, now bring back the screen with a preview of the video on it. At the bottom of this screen you'll see a button called Save and Encode – press it.

Save .avs

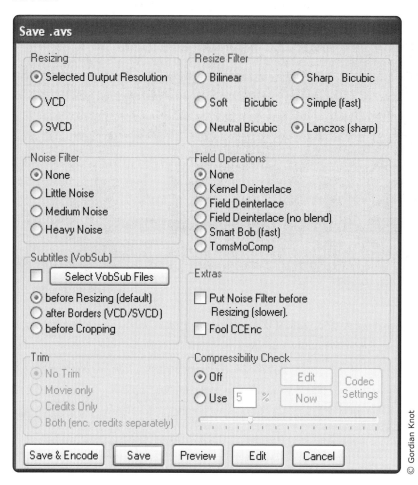

A new window called Save .avs appears. If you're using subtitles this is the chance to make sure the subtitles are included (see the next section on adding subtitles). If you don't have any subtitles simply press 'Save and Encode'. You'll now be asked to give your final compressed film a name and select a folder.

A new encoding box appears – in this example we're going to use the simple DivX 3.11 codec.

Gordian Knot – DivX Encoding Control panel, DivX 3

© Gordian Knot

In the middle of the screen you'll see two tabs called Audio 1 and Audio 2 – the stereo sound channels. Select Audio 1. A series of options now appear. You need to tell Gordian Knot where to find the audio files you've been ripping. Press the Select button and you'll see that inside the main folder, where you've been saving the D2V files, that there are some weird looking file names, usually called something like AC3T01 – open this file.

Now look to the lower right side of this Encoding Panel and you'll see that the audio file has been loaded up. You'll also see that you can click on an option called MP3 custom parameters. A series of complex looking 'Transcoding Parameters' will now load up. Use them all.

Gordian Knot – DivX Encoding Control panel, Audio 1

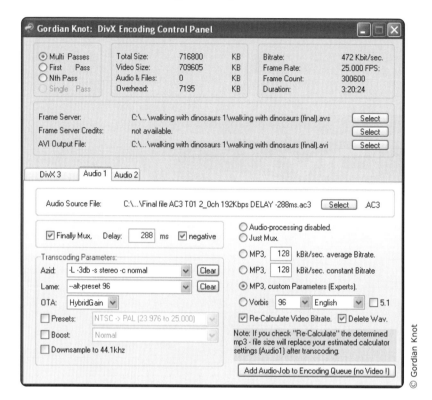

Go to the tab marked Audio 2 and repeat this process. Once you've done this, you've successfully loaded up your audio files.

The last step is to go back to the DivX 3 tab and press a button called Add Job To Encoding Queue.

Time to wander off to bed! Overnight, Gordian Knot will now juggle together all the programs and start the compression process. When you come back the next morning you should have a new small, compressed movie file ready to be transferred to your hard drive.

Using the DivX5 codec

If you want to use DivX 5, the procedure is almost exactly the same except that you'll also be offered the chance to set the parameters by which DivX will encode the files. One of the tabs will now say DivX 5. Click on this and you'll see two buttons in the section called DivX 5 Codec Settings: First Pass and Nth Pass. These options dictate how many times the DivX codec works on your compression.

Click the box entitled First Pass. A new screen will now appear showing the settings for DivX.

DivX – Settings

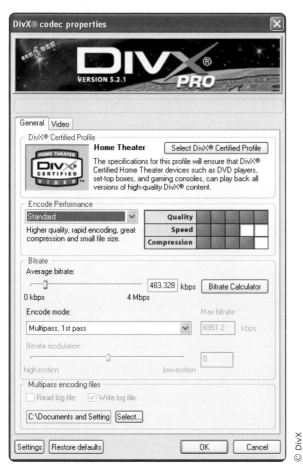

For most of the time you can just click 'OK' and leave the settings as standard (with average bit rate at 483 and encode mode multipass, '1st pass' selected). Gordian Knot will now take you back to the previous box, called Encoding Control Panel. Press on the Nth pass button and repeat the process with the same settings as before.

Remember to click on the Audio 1 and Audio 2 boxes, selecting the right audio file and choosing MP3 Custom parameters as described above. With both the audio and video encoding settings ready, it's time to go back to the DivX 5 tab and select 'Add Job to Encoding Queue'. Gordian Knot will now go away and start compressing the file.

> **note**: If you want you can follow the same procedure to encode using XviD as an alternative codec.

Adding subtitles before you compress the files

If, like me, you're a bit of a foreign language junkie you'll also want the subtitles that came with the DVD. In my experience watching Korean action flicks when you don't understand the language is a bit tricky, though highly amusing. Gordian Knot has a solution, as you'd expect.

With the DVD still loaded in the machine go to the main Gordian Knot page and look at the top of the page for a tab called Subtitles/Chapters.

Gordian Knot – Subtitles / Chapters

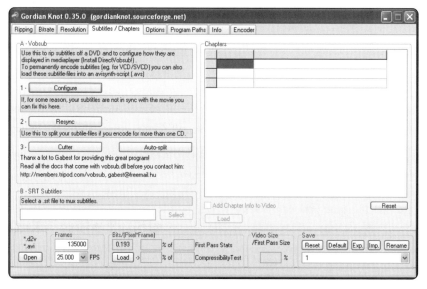

A rather complicated page full of detail now emerges. Ignore most of it and go straight to the button that says Configure.

Filter: VobSub

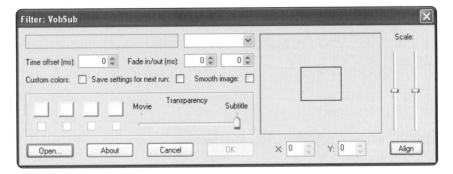

A new screen opens up called Filter:VobSub. Click on the 'Open' button and then navigate your way to the DVD player (usually given the initial D: by the XP operating system). You'll now see two sets of files on your DVD: the video (TS) files and the audio (TS) files.

Double click on the Video box and you'll see a blank folder. Don't panic if you're expecting to see a bunch of subtitle files. Look at the bottom of this 'Open' box and you'll see that it's been told to only look for VobSub files, of which there aren't any. But if you click on this 'Files of Type' section you can also tell it to look for IFO files that can eventually be made into the necessary VobSub files. You'll now see a list of IFO files – pick the one which controls the main feature (usually the biggest file in terms of data size).

The program will now prompt you for a place to save the subtitles file that it's about to create – in my experience, choose the same folder you've saved the ripped files to.

You'll now see yet another box open up.

Select PGC

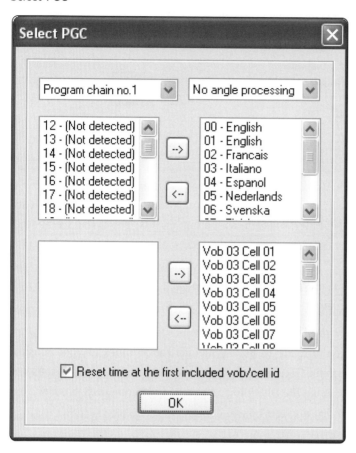

This is entitled Select PGC – it lists all the potential languages the disc has subtitles for. Using the little arrow keys in the middle send all the languages you don't want back to the left column so that you have only one language left (English in this case). Once you have that one language left press 'OK' and the program will now rip out the subtitles, prepare them (it's called *indexing*) and then save them as a text file in the chosen folder. This takes quite some time so be prepared to brew many a cup of tea!

Filter – VobSub

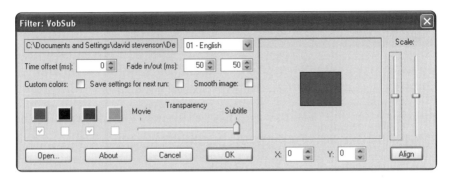

Eventually, when the program has finished preparing your subtitles, you'll see a new screen entitled Filter: VobSub, with a big red box. You'll also see another small text box with the chosen language inside – in this case 01-English – plus some other options that include the transparency of the subtitles. You can now play with the colour of your subtitles by choosing 'Custom Colours' and align the position of the subtitles.

When you're ready press 'OK' to save the resulting subtitle text files to your chosen box and you're ready to go. To include the subtitles in your finished product, go to the main Gordian Knot Bitrate screen. Open up your project as described below (open up the D2V file and choose your compression codec) and then wait for the video preview box to come up.

Once you've made your various tweaks (for framerate, quality and so on) hit 'Save and Encode' on the Video preview box. A new box called Save .avs will appear. Midway down the left hand side of this screen you'll see an option entitled Subtitles VobSub. Tick the Select VobSub box, navigate through to where you saved the subtitle file, and select it (if you see two files by chance there pick the largest one in file size terms). Keep the box 'Before Resizing' ticked. Now go to the bottom of this particular window and 'Preview' the subtitles on the film. A small Windows Media Player will now open up and you'll see your subtitles previewed against the video. If you don't like the colour or position you can go back into the VobSub program and edit the subtitle files.

Step Five - Compressing your whole DVD to a single 1GB file

ratDVD

How about taking your film DVD and compressing it to hell so that it turns into a 1GB file that also includes all the extras and menus built into a DVD?

This is the promise of ratDVD; a great piece of freeware that promises to be the ultimate film DVD back-up solution. It's a better option for backing up films than Gordian Knots compression because you get to keep all the extras. It's also a better option than DVD Shrink for back-up enthusiasts because your end file is relatively small - at around 1GB - and is so small you could store upwards of 80 film DVD back-ups on one small 80GB external drive. And what's more you can still play these compressed DVDs - the ratDVD files - on your PC using Windows Media Player, plus you can also use the software to uncompress the ratDVD file and turn it back into a full, working DVD title.

Program profile – ratDVD	
User	Experienced
What it is	A free program that rips copyright protected movies onto your hard drive and then shrinks them into small files that also have all the DVD extras
Why bother	It backs up all DVD features into a small file which then be uncompressed and burnt again
Source	www.afterdawn.com
Difficulty	Easy
How long will it take to master	Ten minutes

You can get the software from the www.afterdawn.com site or direct from www.ratdvd.dk.

- Download and install as normal.
- Insert a movie in your DVD drive and then open up the program.
- ratDVD does not come with its own decryption software - although there is apparently a third party hack available called RAT Decrypt. This means

you can't use ratDVD to rip the film straight from the DVD if it has any form of copyright protection installed, and you'll still need DVD Decrypter to get the encrypted files off your DVD and onto your PC. Open up DVD Decrypter and go straight to Mode. Choose 'File' – this downloads all the key files from the DVD onto your hard drive. In 'Destination', set the folder you want to store the files in, and then click on the main DVD to hard drive icon. 30 minutes later your files are ready to be converted using ratDVD.

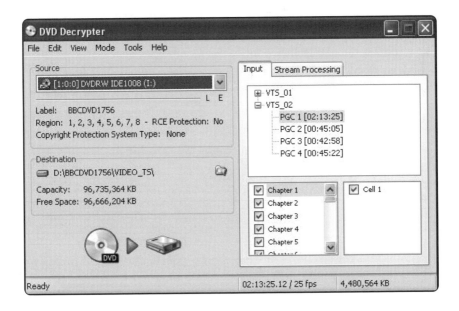

- Time to go back to ratDVD. Click on the 'Open files' folder at the top left side of the main page. Now locate your ripped files. Now open the main menu file called Video TS.IFO.

ratDVD – DVD to ratDVD conversion

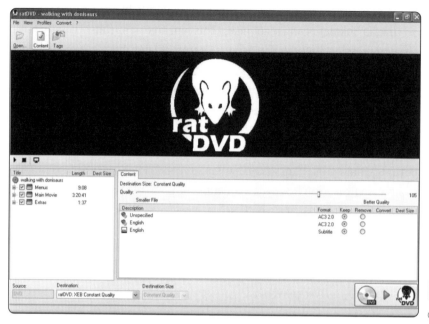

- You'll see all the various elements of your ripped files including the subtitles, the extras and audio tracks. You can now click on any of these options and click Remove or Keep. Unless your DVD folder is huge (close to 9GB) I'd keep most of the extras and options, only getting rid of any useless foreign languages.

- You'll now notice a slider bar under the 'Content' heading. It's called Destination Size and this lets you choose the eventual size of the files. Slide it close to 50 and you'll get a file under 1GB but with lousy quality, while if you slide it close to 125 you'll get superb quality, but the file will be about 2GB. In my experience a setting of around 90 is fine.

- That's it! You're ready to compress. Click on the DVD to ratDVD symbol at the bottom right side of the page. The software is slow and will take at least 5-6 hours to convert, so I'd leave it running overnight.

- When it's finished you'll see a single, hugely compressed ratDVD. To check it's OK, double click on the file and play it within ratDVD – hit the play icon in the middle left hand side of the screen.

5 – Digital Video

> **Tip**: Very occasionally ratDVD may save the final file in a file type that no media player can recognise. This means it's finished all the compression but the file type is incorrect and most media players won't be able to see the file. Don't worry – just right click on the final file, and select rename. Keep the main name but simply add .avi at the end of the file. The file will now read properly in all media players.

- You can also reconvert the ratDVD file back into a full set of DVD files for burning. Simply open up ratDVD again, and use 'Files' to select the ratDVD compressed file. You'll see all the details of your file come up now, and at the bottom of the screen you'll see a new button with 'ratDVD to DVD'. Click on this button and leave overnight. At the end you'll have all the original files decompressed, ready to be burnt onto a DVD.

ratDVD – ratDVD to DVD conversion

In this chapter we learnt how to play video on a PC, and also about all the weird types of video file. You read about (and promptly forgot!) odd names like XviD – don't worry, there's no test at the end of this. You also learnt how to back-up DVDs. You're getting pretty good at this digital video lark by now. But you can't consider yourself an expert just yet. For that, you have to get through the next chapter, which explains how to download video files on the internet. So, without further ado, let's get onto file sharing networks.

File Sharing Networks

Content galore, pity about the law

Imagine a series of networks based around the internet where you could download millions of different files – music, film, software – all for free. It's no socialist utopia, but a grim reality for the entertainment and software industries – global online communities like BitTorrent, eDonkey and Gnutella are the decentralised hubs of a new world of free digital content, shared over what are called peer-to-peer networks (P2P). It's all great news for internet users, but terrible news for the corporations that own the digital content!

Every day, tens of millions of users download and upload files for every imaginable type of content. Some of that content is free of copyright and entirely legal, but most of it is copyrighted material that legally belongs to someone else and is being shared in an unauthorised fashion, globally.

Before we go any further, it's important to say that file sharing technology, and the networks themselves, are not necessarily illegal. Far from it, in fact. Many big software houses, like Microsoft, are hard at work building clever tools into their operating systems that allow perfectly legal file sharing, between work colleagues, for instance, or between offices.

The problem with the file sharing networks featured in this chapter is that much, if not most, of the content on these networks is illegal – very simply, it shouldn't be there! We'll talk about the legal position of these networks and the files on them in a later chapter, but the bottom line is simple: share files over networks like BitTorrent and technically, in most countries, you're breaking the law. That doesn't mean you'll get caught or that there aren't some arguments in favour of file sharing copyrighted material, but it's the reality. It's illegal but millions of people use the networks anyway. In fact, tens of millions!

It's also important that you realise that many of these networks are full of:

1. **Fakes** – bad and corrupted files that are next to useless and frequently placed on the networks by operatives of the entertainment industry.
2. Files full of **viruses** and software full of **spyware**. We'll talk about this very real threat in the next section.
3. **Hackers** desperate to get past your security and corrupt your system. One quick reassurance on this theme though: if you follow one simple rule you should be able to stop this particular threat dead in its tracks. Apologies for

repeating this yet again, but *never, ever, visit a file sharing network unless you have a fully operational firewall.*

> **note**: So let's be very clear before we carry on: before you use a file sharing network understand clearly that your downloads (and uploads) might well be potentially illegal and not without hazard or danger to your own computer.

But there is a big upside that goes with these risks, namely that virtually any film or album you've ever wanted is probably on one of the networks, for free. All you have to do is to work out which network to use and how to get the file without landing yourself in gaol or spending hours cleaning up your PC.

In this chapter you'll:

1. Learn the principle behind file sharing networks.
2. Learn how to work out which network to use.
3. Discover the wonders of BitTorrent.
4. Work out how to use the slightly more complicated eMule network.

What exactly is a file sharing network?

Most commentators use a number of interchangeable terms to describe the burgeoning world of file sharing. You'll see talk of peer-to-peer networks (P2P), file sharing and even decentralised, distributed computing – for the purposes of this chapter though they're all the same thing. Whatever technology or network you end up using, they're all probably built on the same foundations.

Version 1: FTP

In the time before P2P, something called file transfer protocol (or FTP) was the very first incarnation of file sharing and involved you going on the internet to find a file and then requesting to download that file from a central computer, or server. That server almost certainly belonged to someone else and you either had to have their permission or pay for the privilege. Even worse, if you and a thousand other people converged on the same server at the same time and tried to download the same file, it would have probably keeled over unless it was a very big computer with loads of bandwidth.

Version 2: P2P

Napster style. Modern file sharing networks try to do away with all this downloading from a central server nonsense by distributing the file over many different computers on a network (i.e. the internet). The problem is then how to organise that distribution and then find the files.

Step forward a certain Shaun Fanning in late 1999.

Apparently all that Mr Fanning wanted was a nice and simple way of finding decent music over the internet without having to trawl through instant messaging systems or web search engines like Lycos and Google to find someone who had posted some music on their website. Clever young man that he was, he worked out a series of technological innovations that led to something called Napster. Although it's worth pointing out that other systems like IRC, Usenet and Hotline had been around a little earlier with similar goals and objectives, Fanning's genius was to develop his own, specially designed software that centralised the indexing of all the music tracks on his network.

Here's how it worked.

If you were looking for a digital copy of a music track, you'd transmit a search request to the Napster server, where the software would conduct a text search of the centralised index for matching files, and the search results would then be transmitted to the requesting user. If the results showed that another Napster user was logged on to the Napster server and offering to share the requested music, the requesting user could then connect directly with the offering user and download the music file.

Big, legal mistake!

Why? Fanning's servers were obviously owned by his company and it was clear to all and sundry (except his lawyers) that they were indexing content that shouldn't be there and, what was worse, Napster's management clearly knew what was going on. Within two years his network had been shut down and Fanning was forced to close the Napster servers and sell up – ironically to a media giant called Bertelsmann who reopened the service a few years later on a legal, paid-for basis.

Version 3: The new generation of P2P

Unfortunately for the music industry – and later, the film companies – the genie was now well and truly out of the bottle as other developers started work on a new generation of networks that were much more decentralised and less vulnerable to legal assault. In these decentralised indexing networks, each user maintains an index of only those files that the user wishes to make available to other network users. Under this model, the software broadcasts a search request to all the computers on the network and a search of the individual index files is conducted, with the collective results routed back to the requesting computer. This clever model is used by software (like Morpheus) that is based on a set of open source code called Gnutella, which allows modification of the software, subject to some restrictions.

A rival soon appeared based on something called the *supernode* model, in which a number of select computers on the network are designated as indexing servers. The user looking to search for a file connects with the most easily accessible supernode, which conducts the search of its index and supplies the user with the results. Any computer on the network could function as a supernode if it met the technical requirements, such as processing speed. This technical architecture was developed by a bunch of programmers behind a company called Kazaa, who in turn made pots of money by building a hugely popular network called Grokster.

But the techie types were still on edge, as the anti-piracy watchdogs could still attack the companies that operated these supernodes and marketed the network. If you really want to stay under the anti-piracy radar you'd need an even more decentralised model with technology that could cope with all these massive, new film files emerging.

Cue Bram Cohen's *BitTorrent* system, which debuted back in 2002.

The world according to BitTorrent is one based on swarms of computers, all acting as servers. The basic principle behind this kind of network is that all the files (typically films) are broken down into much smaller fragments (under a half MB each) that are distributed to the peers (you and me) pretty much randomly and then reassembled on a requesting machine. Each peer takes advantage of the best connections to the missing pieces while providing an upload connection to the pieces it already has.

Without getting too technical, its genius is that it breaks huge files down into manageable packets that can then be distributed in a swarm-like fashion around the internet. This overcomes another key worry; that of bottlenecks emerging (bandwidth, the measure of capacity on the internet is not unlimited after all). It does this by cleverly turning all that increased demand into yet more bandwidth as your computer is turned into yet another server that can help new entrants.

Which networks to use?

If it all sounds a little too good to be true, it is. It costs money to develop many of these technologies and market the networks, money that's certainly not going to come from all those file sharers who are only using the network because it's free.

So, companies like Kazaa turned to novel ways of making money. Advertising soon made an appearance on its network, usually in a highly annoying and obtrusive fashion. The owners of these networks also worked out that the software that was installed on your computer could also be loaded with wonderful little extras (spyware) that could be either slipped in without any mention, or referred to as a useful adjunct to your internet search and shopping experience (spyware again).

In essence, that means many of these networks are infected with horrible spyware, like Gator, that purports to help you but is in fact spying on you and making your life difficult. Luckily, the net is also full of well-meaning users who spend a lot of their time checking whether the software that comes with these network installation packs is infected or clean.

Step forward the excellent folks at SpywareInfo who've compiled a handy list of all the networks and the accompanying software that's infected. You can see the full list at www.spywareinfo.com/articles/p2p/ but in the table below I've summarised the chief villains.

File sharing networks – ones to avoid		
Audiogalaxy	iMesh	FreeWire
BitTorrent Ultra	Warez P2P	LimeWire before 2004
Bearshare (free version)	Morpheus before version 4.9.2	Grokster
MediaSeek	Edonkey prior to version 1.2	

Source: Spywareinfo.com

My advice is simple – don't bother using any network or software client that's infected. That means avoiding Grokster, LimeWire, Bearware and early clients of Edonkey.

Luckily there are also an equally large number of networks and software clients that are clean – here's the SpywareInfo list.

File sharing networks – the clean ones		
WinMX	Shareaza	eMule
Soulseek	Direct2Connect	BitComet, Torrent Storm, BitTornado, and TorrentSEarch, Azureus (all BitTorrent clients)
Filetopia	Morpheus (after version 4.9.2)	Ares Lite

Source: Spywareinfo.com

Choosing which of these clean networks and clients to use boils down to three simple factors:

1. How easy is the software to use?

   The bottom line here is that it's really not worth touching a client (software program) that's complicated to use. That rules out the various IRC-based clients, of which the best is something called AutoXDCC that lets you access a fantastic treasure trove of content, but it's fiddly and difficult to use unless you know lots about internet relay chat. It's a similar story at the excellent Direct Connect – great client, loads of users on the network, but best avoided by novices.

2. How much useful content is available on the network?

   The Ares network looks hugely promising and is very easy to use, but at the moment the number of users is still fairly limited. It's a similar story with Filetopia and Soulseek.

3. How fast are the downloads?

   The various BitTorrent clients win hands down on this score, while eMule is a veritable tortoise by comparison.

That leaves us with just four serious contenders: Morpheus, WinMX, eMule and the various BitTorrent clients.

In this section we're going to concentrate on just two networks: BitTorrent and the eMule client on eDonkey.

Why not Morpheus or WinMX?

File sharing network Morpheus used to be the clear market leader, but the number of files available has dropped dramatically in recent years and its also become popular with the wrong sorts, who in turn have placed loads of dodgy files – viruses and malware – on the network. Its past popularity also made it a major target of the anti-piracy authorities and its owner is facing a string of legal actions designed to shut it down.

To be fair, Morpheus has many things going for it – it uses the excellent Gnutella network and is incredibly easy to use – but I'd avoid it simply because there are better, faster, easier to use alternatives out there, although if you're really determined to use it we have included a small section (starting on page

269) on Morpheus basics, plus a quick look at LimeWire, another, once popular, network that's worth a look if you don't mind loading up some spyware.

WinMX was, and quite possibly still is, a great network, but back in September 2005 it appeared that it had shut down. By the time you read this book, WinMX has probably come back online via some clever work-around, but its prospects don't look great. It's also not quite as easy to use as the BitTorrent and eMule clients and its best content is pretty much limited to music and software.

BitTorrent vs eDonkey

The BitTorrent and eDonkey networks (only use the eMule client) are both sturdy, reliable survivors, jam-packed full of content with easy to use clients that are free of spyware.

Table 6.1: Comparison of BitTorrent and eDonkey networks

Network	Advantages	Disadvantages
BitTorrent	Downloading files is incredibly quick. There's an astonishing quantity of files.	A little fiddly to use. There's a vast amount of content available, but it's usually only the most up-to-date releases. It's not great for esoteric and slightly less mainstream content.
eDonkey/eMule	Slightly easier to use. Boasts an astonishing number of files catering to an enormous variety of tastes. It's the place to go for alternative tastes and old films and music.	Downloading can take much, much longer as it operates a queuing system. Heavily patronised by non-English speakers and especially Germans. That means it can ack depth when it comes to Anglo-Saxon music and film.

> **note**: My advice is simple – use both networks!

BitTorrent – the new kid on the block

In little more than three years BitTorrent has emerged as the dominant P2P client and the new Public Enemy Number One of the film and music industry. BitTorrent is unique in several respects:

- **It's a smart network**

    Whereas most file sharing systems allow you to download and upload files from one source at a time, BitTorrent cleverly takes chunks of data simultaneously from anyone online who has it. And where P2P used to be about sharing individual music tracks, the combined power of BitTorrent and broadband means you can download the entire Beatles catalogue or an entire TV series as a single compressed file in just a few hours.

- **No central servers as targets**

    The problem for BitTorrent's opponents is that there are no central servers storing the files or allocating their distribution, thereby depriving them of an easy, legal route to shutting down the network. BitTorrent also dares to have an agenda. It has an anti-spam and anti-porn policy and, touchingly, allows users to make one-click donations to the PayPal account of Bram Cohen, its 29-year-old designer, who has (so far) resisted the temptation to sell out to big business before the lawyers close in.

- **Share bandwidth and get more**

    The network's agenda positively discriminates against 'leechers' – those who download files and give nothing in return. By tracking your download and upload speeds, you receive progressively slower downloads the more you tweak your settings to restrict upload bandwidth. This leads to a genuine community spirit that does a better job of policing itself than other P2P networks.

Luckily for its opponents – and there are many – BitTorrent isn't as easy to use as some other networks. It does have some other downsides of which perhaps the biggest is that its content, though incredibly broad, is not actually very deep. What this means is that its content is usually very contemporary – the latest films, TV shows and albums – but if you're after much more specialised stuff it'll probably not be available. Other networks, like eMule, are probably a better bet for a much deeper range of back material (we'll talk about this network next) but if you're after the latest thing then BitTorrent is for you.

BitTorrent lingo

Many users are put off using this fantastic network by the strange language that is used by the network's varied developers. Don't be. BitTorrent does take a bit of getting used to but it is, in reality, a doddle to use.

First, lets get our head around some of the terms used.

- **Seeder**

  This is simply someone who has the complete file and lets others download it using BitTorrent. But a seeder is also someone who leaves the file going after they've finished downloading it.

- **Leecher**

  This is someone who is downloading a file but hasn't finished it yet.

- **Reseed**

  This is used to describe the process where someone seeds a torrent again after they'd already finished and closed it earlier. This is done when a torrent has many leechers, but no seeder. This is a great way to help out the community of BitTorrent users as others can now finish their download and then seed again.

- **Tracker**

  An application run by the person who runs a BitTorrent website. It controls the up/downloads of the torrents which use the tracker's web address or url.

- **Peer**

  Someone who is uploading and/or downloading a torrent. Generally a peer does not have the complete file, otherwise it would be called a seed. Some people also refer to peers as leechers, to distinguish them from those 'generous' folks who have completed their download and continue to leave the client running and acting as a seed.

- Torrent

    This is the small data file you receive from the web server (the one that ends in .torrent). It contains the information about the data you want to download, not the data itself – this is what's sent to your computer when you click on a download link on a website. You can also save the torrent file to your local system, and then click on it to open the BitTorrent download. This is useful if you want to be able to re-open the torrent later on without having to find the link again.

- Swarm

    The group of machines that are collectively connected for a particular file. For example, if you start a BitTorrent software program like BitComet (see below) and it tells you that you're connected to 10 peers and 3 seeds, then the swarm consists of you and those 13 other people.

Putting together all the jargon

You should now be able to build a picture in your mind of how BitTorrent works.

The main building block is that there are people (seeders) who upload the digital content to a website to which they apply a tracker that controls the traffic in the file. These files are then scanned by clever BitTorrent search engines that are a bit like Google – they send electronic spiders over the web to search for BitTorrents and then present the compiled list of available files to other users who could be either seeders or leechers. These seeders and leechers then get to work, downloading the BitTorrent tracker file (which tells them where the file is and how to use it) from the website identified by the search engines, pushing the content out over the web in a matter of minutes. These users in turn act as suppliers of content to leechers who simply download content without ever actually providing content for further redistribution.

What you'll need

To run BitTorrent requires a number of key tools:

1. You need a clever search engine. Many have been shut down but a few major sites like www.torrentspy.com and www.isohunt.com keep going, barring the inevitable legal threats from the anti-piracy police. I'll show you how to search for content below.

2. You need a program that controls your uploading and downloading. There are plenty of programs available, ranging from the original BitTorrent program itself all the way through to highly optimised programs/clients such as BitComet that are very easy to use.

3. You might also decide to seed content. Again you might need a special bit of software – in this case we'd use a very popular little program called Make Torrent. This is a fairly complicated procedure for file sharing novices and we won't go into it here – if you want to master seeding a download, visit krypt.dyndns.org:81/torrent/maketorrent/ where you'll also find a FAQ and User guide.

How to search for BitTorrent content

Let's see if we can download the latest episode of a popular US TV show, *West Wing*.

We're looking for this kind of content because TV shows make for reasonably small downloads – 350MB – compared to film files which start at 600MB and can end up being as big as 1.5GB in size.

TorrentSpy

Our first stop is the website of TorrentSpy – www.torrentspy.com.

TorrentSpy – Main page

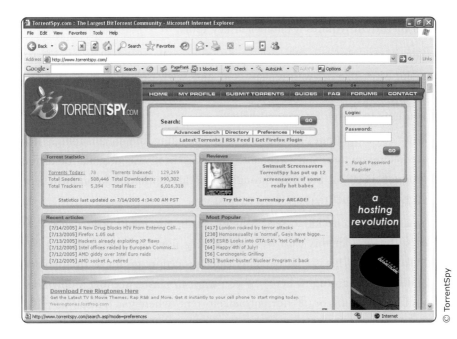

There's absolutely nothing complicated about TorrentSpy. It's a straightforward search website that looks for all kinds of BitTorrent content – legal and illegal (although the website does say that it conforms to the US copyright laws by condemning illegal, copyrighted files).

On this big, very long page you'll see a long list of reviews, news and user comments and stats, but the key box is at the top – Search.

If you want to run a very simple search you could simply type in 'West Wing', but lets be a little more specific and double click on the Advanced Search link.

TorrentSpy – Search page

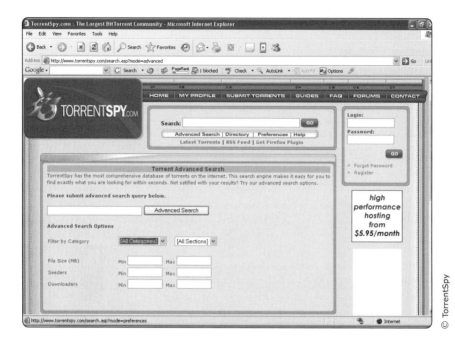

We now see a number of filters that allow you to be more specific with your search. As it's a TV show we're looking for, click on the 'Filter By Category' drop-down and select TV. You can also specify the file sizes and the number of seeders and leechers using this file, but lets just keep it simple. Type in 'West Wing' in the main search box.

You'll now see the results of your search – hopefully, a long list of available files. These can range from huge files that contain all the episodes in a series through to individual episodes.

You'll also be told the size of the file – the bigger it is the longer it takes to download – and the number of seeders and leechers using the file (the more the better). You'll also be told the number of files within the file you're downloading – this can range from 1 (a simple AVI video file), through to more than 50 if a zip/archive tool such as WinRAR is being used.

Lastly, there's a very useful Health column with coloured indicators – green means the file is easy to access and in good health, while red means it's difficult to access and is probably not worth the bother.

Click on the healthiest file, which in our case is the series 6 collection – this means the file contains all series 6 episodes. Click on the file.

TorrentSpy – Results

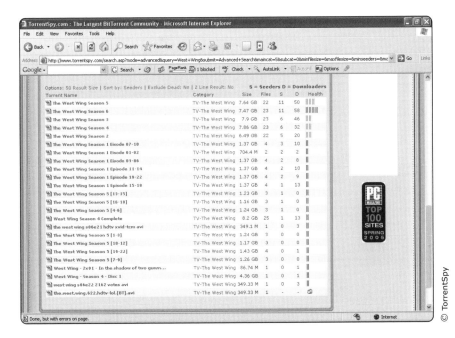

You'll now be given the opportunity to download the torrent file – you'll also be given a lot more information about the file. You'll be able to see when the file was first added, exactly how many seeders and downloaders there are plus, further down the page, you may also find a section of comments by users. This is incredibly useful as it tells you if the file is a fake – someone is bound to announce this, usually in big capital letters followed by an expletive or two – or if the file is corrupted in some way. Double click on the Download Torrent file. You'll now be asked to save the file. Create a folder inside My Documents called My BitTorrent files and save the file in this newly created folder.

TorrentSpy – Downloading

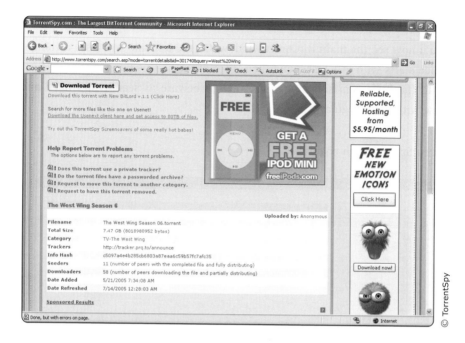

We now have a BitTorrent file that will tell us where to download the file from, but at the moment it's completely useless to us as it stands – we have no program that tells our computer what to do with it. We need to download a BitTorrent client program.

BitTorrent client programs

There are an enormous number of BitTorrent clients freely available and many of them can do some pretty amazing things, but for simplicity I recommend getting BitComet.

To find BitComet go to www.download.com and search for 'BitComet', or go direct to the BitComet website at www.bitcomet.com.

Download the file as normal, letting it install an icon on your start menu and desktop. It will then prompt you to run the program – click 'Accept'.

Almost immediately your firewall program will tell you that BitComet is trying to install itself and connect to the web. Allow the BitComet client software to connect to the internet (remembering to tick the 'Always' box if your firewall software allows, so you don't have to keep telling the firewall to authorise the client).

You'll now see the main BitComet window. On the left side you'll see a column which starts with 'All Tasks' and then 'Torrent Sites'. This latter section is a list of all the BitTorrent search engines, though beware as search sites open and close with great regularity.

TorrentSpy − Search engines

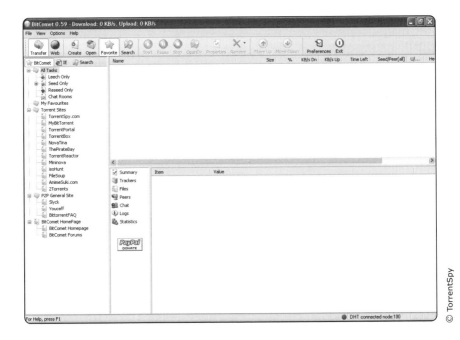

Now go back to where you saved the *West Wing* torrent file. Double click on this tiny little BitTorrent file. A new window opens up giving you details of this torrent file using BitComet.

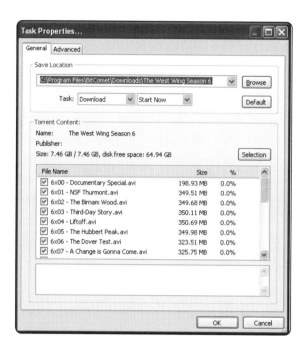

Just press 'OK' and BitComet goes away and starts downloading your files. The program will also minimise itself. If you want to check on the progress of the download at any time you can go to the bottom right side of your screen where you'll see the icon for BitComet. Double click on this icon and BitComet will appear again. You should now be able to see your download file with a green arrow pointing down (that means it's downloading) and some columns that will tell you the size of the file, and the download and upload speed. You'll also be told how long it will take to download the file at the current speed.

It will probably take a few minutes for BitComet to contact all the seeders out there and arrange the download (or swarm) so don't be surprised if the speed starts very slow and then builds up.

If you want to change anything about the download – deleting it for instance – just right click on the file name. You'll see a number of options that include 'Remove'. Once you've finished downloading your file you can choose to Remove either the 'Task' (stops it being seeded) or the files as well.

If you do stop BitComet and then want to restart it again, you'll have to restart your downloads. To do this simply right click on the files you want to restart downloading and select Start.

In the later versions of this client there's also a button on the top screen of the program that will let you preview the files you're downloading.

Firewall problems

If you're using a software-based firewall, you need to enable incoming connections so that they can deal with BitComet and other clients. If you're still using only the most primitive of firewall programs – the one supplied free with Windows XP – you'll need to configure the software to let BitComet through safely. This means you've got to open up some of your communication channels – or ports – for BitComet using Windows XP firewall. To do this:

- Open the Network Connections folder (Start > Control Panel > Network and Internet Connections > Network Connections).
- Click the shared connection or the internet connection that is protected by Internet Connection Firewall, and then, under Tasks, click Change settings of this connection.
- On the Advanced tab, click Settings. For each port you wish to forward, (i.e. 6881, 6882... 6889) do the following: On the Services tab, click Add, and enter all of the following information; in Description of service, type an easily recognised name for the service, such as 'BitComet'. In Name or IP address of the computer hosting this service on your network, enter 127.0.0.1 (this means 'the local machine'). In both External and Internal port number for this service, enter the port number, e.g. 6881. Select TCP, then OK.

If you're running proper firewall programs like Sygate or ZoneAlarm, you should be prompted to open up these channels by the software.

Tips and tricks with BitTorrent

- Change the location where BitComet downloads your files – go to Options, at the top of the screen and then select Preferences. Here you'll see all the various options for running BitComet – in the left column you'll see 'Task' with a small folder symbol next to it. Click on this and then double click on the first option in Download that says Default Download path. You can now change the location of your shared and downloaded files. I use a folder called My Shared Files, which all my peer-to-peer programs download files to.

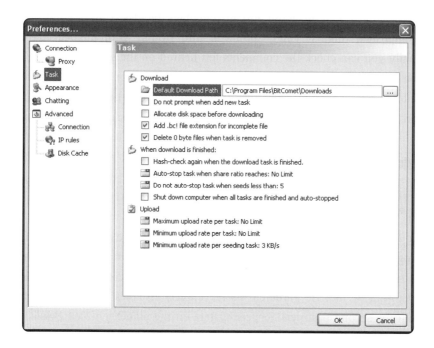

- If there are no seeders there are no files to download. Look for files with plenty of seeders and downloader's/leechers.
- Really do look at the comments on a search engine like TorrentSpy. If someone says 'Fake', 'Virus' or 'Crap' use your common sense and avoid it. There's also a remorseless logic to popular files – by and large the more seeders/downloaders there are, the more likely that the file is healthy and uncorrupted. If the health indicator in TorrentSpy is red and there are not many people using the file, investigate and probably avoid.
- You'll also encounter some rather weird file formats. The most common are .rar, .cue and .bin. Typically you'll download what you think is a big file inside a folder. When it's complete you look inside the folder and see dozens of different files with these initials at the end. Don't panic! They're all just ways of compressing and managing data files – our earlier chapter about video talked about opening up .RAR files.
- It's common sense, and we've already laboured the point in an earlier chapter, but always run a virus check on all the files you download. Do not go blundering into a file hoping it will be OK – scan it first.

- Be particularly careful about foreign language movies. Unless you're an accomplished linguist you'll soon run into trouble with dubbed films. Usually you can check the comments left behind on the search engines or even inspect the name of the file (it might say 'Italian' for instance – which is a bit of a giveaway for Italian language). These might be dubbed with no subtitles. Unless you speak Italian it's likely to be useless.

eDonkey/eMule

What is eMule?

Back in 2002, a clever chap called Merkur decided that the existing peer-to-peer programs were dreadful, especially the early incarnation of a network called eDonkey. As is the way with these techy types he was convinced he could do better. And he did. He gathered his (developer) disciples around him and they built the second best peer-to-peer system on the planet, entitled eMule – according to the site it was so called because a Mule was 'somehow similar to a donkey'. It was built on the existing eDonkey infrastructure but it boasted lots of new features and a user-friendly interface and it's completely free of any spyware or malware.

eMule is a great P2P tool. Compared to its rivals it boasts a number of big advantages:

- It's hugely reliable and its myriad of servers located all around the world make sure that the network is up and running no matter what. That reliability also explains why it's recently become a target for the anti-piracy police.

- Its main user interface is very easy to use and understand and has loads of useful features.

- The network itself boasts huge amounts of content. This is its biggest advantage over BitTorrent – eMule's users are much more likely to be cultivated, intelligent, sophisticated (aka non-American) types who like world movies or strange unheard-of bands.

- eMule makes use of a number of other networks including ED2K, the Source Exchange, and Kad.

- You can preview content – that means you can see a video file even before it's finished downloading.
- eMule also boasts another nice feature in its Queue and Credit system. This makes sure that everyone in a queue will get the file they want by promoting those that upload back to the network.

How to install and use eMule

Download the eMule software from its website – www.eMule-project.net.

- On the home page you'll see a link to 'Download'. Click to download the latest release of the software – currently version 0.47a.
- After the main installation, run the program for the first time. You'll come to the configuration process. First it will ask you for a user name (make one up and don't use your own name, ever!) and then you'll be asked to check the communication ports used to run the program. You'll spin through various other options – you can accept all the defaults for these.

Eventually you finish the configuration process and you'll find yourself on the main page of eMule.

eMule – Main page

Don't start using the program yet. In the main window you'll see a list of servers that you can connect to using eMule. This will probably be a rather short list – and it certainly won't include the most powerful and popular servers. For that we've got to go back online and visit http://ed2k.2x4u.de/. Here you'll find three choices – the connect list, the best servers and all servers. You'll also see the option to add any of these to eMule. In my experience you should go for the Best Servers and Add to eMule – all the addresses of the most popular servers will now be added to your server list.

With your extended server list we're ready to connect to a server. Personally I prefer to connect to the most used servers as these probably have the most content on them. To find this out click on the column in the server list that says Users. The server list will now be filtered on the basis of the most popular first – which usually means one of the Razorback servers, although be aware that the anti-piracy police in the Netherlands have recently shut down some of the bigger Razorback servers.

Now right click on this server and select 'Connect'. After a few seconds you'll see that you're connected in the 'My Info' box in the bottom right corner.

Congratulations – you're on the network!

Before we start looking for files, let's make sure we have selected the right configuration options.

The Downloader's Handbook

You can now tell the program which folders you want to share with it.

- Click on the drives or folders. You should also select one main folder in which you store all your incoming and temporary files. As with BitTorrent I'd reset this to a specific file called something like 'My P2P Shared Files' in My Documents. That way, all your peer-to-peer content goes into one folder, which can be easily checked for viruses. Once you've selected the shared folder, click 'Apply' and then 'OK' and go back to the main screen.

Looking for a file

Time to start looking for a file.

Let's start by seeing if we can download an episode of the *West Wing* again.

- First click on the Search button at the top. This brings us to the main search page. Type in your search term and then hit 'Start'.

eMule – Search page

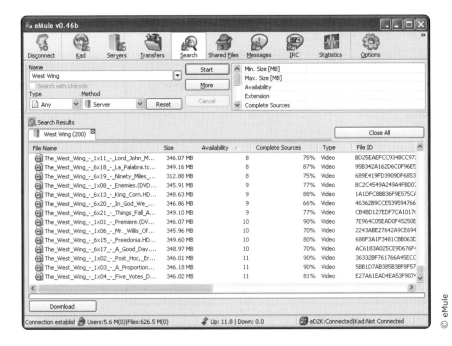

- You'll now see a long list of files of different episodes come up. The important columns are in the middle. One says 'Availability', which tells you how easily you'll get hold of the file (a high number is good especially if it's in blue), while the other says 'Complete Sources'. This last column is hugely important as you really only want files that are complete – what's the point of downloading incomplete files!

- After filtering the files for their completeness, look down the Availability column and see what's complete *and* has plenty of availability (if it's in blue that's good).

- When you've found your file, highlight it, and then click the download button at the bottom (you can also right click the file) and select download.

eMule – Downloading

- To see how the download's going click on the Transfers button at the top. Hopefully you'll see your chosen file in the list there. You'll see another bunch of columns as well. First you'll see the size of the file you're downloading, next the speed of the download when it eventually starts, and then a progress bar. If you see any red in this bar, avoid the file – right click on it and choose 'Cancel download'. Red means that bits of data in the file are unavailable – there's no point in downloading these files as it's not complete. The darker the blue, on the other hand, the better your chance of downloading. You'll also see columns that tell you the amount of people who have the file online (clients with the file) plus your status (it'll probably say 'waiting' at first) plus the amount of data still to be downloaded.
- At the bottom of the screen you'll see a separate box that tells you the number of people who are downloading from you.
- Lastly, you'll also see a couple of little 'i' symbols next to some download files in the transfer section. Most of the time these should be green in colour – they are comments made by other users about the files you're

attempting to download. A lot of them will be in foreign languages but you should easily spot words like 'FAKE' or German terms for excrement (very common with bad files in my experience).

Tips and tricks with using eMule

- Don't pen your eMule in!

    eMule behaves like a server, and that means it wants to accept connections from other peers. If you trust that eMule is relatively secure then you should allow it full access to the internet and not try and use your firewall program to restrict access to the internet. If you use a firewall, like ZoneAlarm, allow connections in both directions – both incoming and outgoing. You can check how penned in your eMule is by looking at the bottom right of the client window: if you are connected to a server then the little icon should have green arrows round the logo.

- The Kad network

    This is an alternate way for clients to find each other rather than using the main ED2K server system (it's a good idea to use both Kad and the ED2K servers). There's not a great deal to say about this, just turn it on and connect it. To enable Kad go to Preferences > Connection and check 'Kad' in the bottom right corner. To connect to Kad should be simple, go to the Kad icon at the top of eMule and the first time press 'Bootstrap', it should just connect.

- Be patient

    You may well be wondering why your program has only downloaded a few Kbytes in the last few hours. The answer is simple – eMule works by downloading off clients and uploading to clients, and although many people do run eMule all the time, they also probably have hundreds of people in a queue waiting to download. And like all queues you have to wait your turn!

Other networks

I make no apologies in this book for concentrating on the two best global P2P networks: BitTorrent (using the BitComet client) and eDonkey (using the eMule client). They're both fantastic, full of content, and relatively easy to use.

But there are some perfectly viable alternatives worth exploring if you don't mind taking on some extra risks. In this section we'll quickly look at just three of them: Morpheus, Bearshare and LimeWire.

Each of these networks have their pros and cons.

Table 6.2: Comparison of Morpheus, BearShare and LimeWire

Network	Advantages	Disadvantages
Morpheus	It's free of spyware in its current version. Has an enormous number of users with a vast amount of content.	The specific target of the anti-piracy cops and that means users are much more likely to be targeted by legal actions. It's very popularity has also made it a target for unscrupulous peddlers in fake files and malware.
BearShare and LimeWire	Well established user base with loads of content. Easy to use interfaces that make downloading very straightforward.	Software clients that are riddled with spyware.

On balance, Morpheus is probably the best of the bunch – although as I write this there are rumours that it's about to close down!

Morpheus

Morpheus used to be the king of P2P networks. Its genius was that it identified high-powered peers – those with fast CPUs and speedy internet connections – and set them up as supernodes, which carry information about the lower-powered peers and help facilitate each search.

In a traditional, pure peer-to-peer network, a query would travel to and from each individual peer, hogging processing power and bandwidth. With Morpheus, queries go to the supernode, which then decides whether to send them down to the lower-powered peers. Because of its size and its use of supernodes, the Morpheus network typically gives you access to a wonderfully extensive, eclectic array of songs and other files.

To download the program go to www.morpheus.com and install as normal.

Morpheus – Front page

Searching for files using Morpheus

Select which kind of media you want by choosing the 'All Types' option, then enter your search criteria and press Search. You can also run multiple searches and downloads at once!

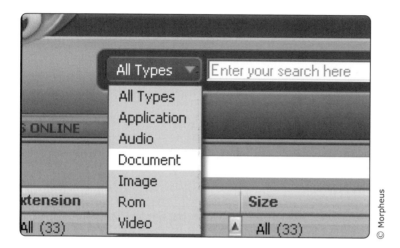

To organise and target your search, you can use the 'Narrow Search' option.

> **Tip**: If you are having difficulty finding the file you want, you should try again later by clicking the Search More button or leave a search result open. Each time someone is connected to the network, their server is scanned for files adding to the new results.

Sharing files

Morpheus includes sharing features that make it easy to find and download content, but you need to make sure you only share the files that you are willing to allow other Morpheus users to be able to download.

Media files can be shared in two ways: within the My Files tab or directly from the Preferences button shown to the right. To share, go to 'Preferences', and then select Folders and select the Add button and then browse your PC and its disks to share individual folders and subfolders. Using the file chooser window, browse to the location of the files that you would like to share, and select Open. Repeat this process to share multiple folders in different locations.

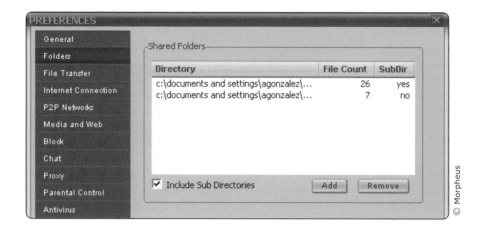

new tools within Morpheus

1. Parental control

The latest incarnation of Morpheus also includes much improved password-protected parental controls. This gives you the ability to add some extra layers of protection, security and preferences to your Morpheus experience.

To access these functions, navigate to 'Preferences' and then select 'Parental Control'. If this is your first time accessing these settings, you should create a password that is then used for all the functions in this area.

2. Proxy

Proxy servers act as intermediaries between internet users to protect their identities while active on the internet. To search for free proxies visit www.proxyblind.org which can then be added to Morpheus by going to Preferences, and then selecting Proxy for anonymous downloading.

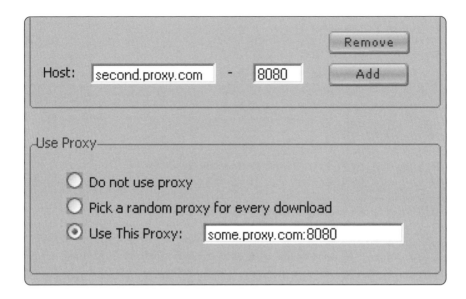

3. Virus scan

You can even set up your virus scanner to automatically scan downloaded files from within Morpheus. Go to Preferences and then Antivirus. Click on Add and locate your virus scanner to automatically scan every completed file you download, adding a measure of security in keeping your computer safe. If you use AVG, the address will probably be something like C:\Program Files\Grisoft\AVG6\avgse.exe while Avast will be C:\Program Files\Alwil Software\Avast4.

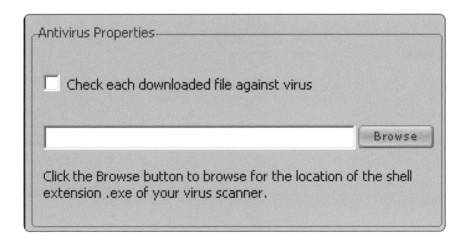

BearShare

Great network, pity about the spyware!

But don't completely ignore BearShare if you're after hard to find material not available on other networks as it does have an enormous amount of content lurking around on its network and may be worth a look if you can put up with the spyware.

To get the installation file go to www.bearshare.com, select Download Now and then hit the Free button to start the install file downloading. Install the program as normal.

Using BearShare

The first time you run BearShare, a set up wizard will open to configure your BearShare:

- The first window will welcome you. Click Next.
- The next window lets you set up preferences. It's best to leave as the defaults and you can always change them later. Click Next.
- Next select your internet connection. Click Next.
- Now you'll be shown where the files will be saved. To change, click Explore or Browse and select a new directory. Click Next.
- The next window will show you which folders are to be shared across the network. It's best to deselect all and then choose which you want one by one. Click Next.
- Click Finish.

A pop-up will come up with tips – these are often useful and worth reading. Now click Close.

Searching for files

Time to start searching...

Click the Search tab at the top left of the window.

Enter some keywords in the box titled Find. Also set the type of file you are looking for (audio, film, image, etc). You can use advanced criteria but this will narrow the search so you are more likely to end up with fewer sources. Click Search.

A list of the searches in progress and completed is kept on the right of the window. To move from one search to the next click on them. You will also be shown the number of hits to sources (x/y) and the status of the search. It's always best to wait till the status reaches 100% before starting a download to make sure the file you choose is the one with the most sources, as there can often be many different versions of the same file each with different numbers of sources.

BearShare – Searching

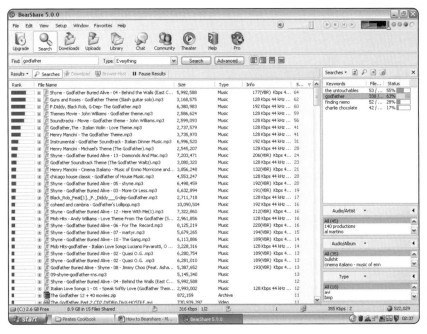

- To start downloading, first check you have the right file. Look at the column on the far left of the main page – it gives you a visual representation of the number sources for each file. They show full when they reach 100 sources. By default the number of sources are shown on the far right column. You can choose which column to be made visible by right-clicking a column header and checking the categories that you want. When downloading choose the hit with most sources for the fastest download.

- Obviously, check the file name in this search box but read the whole list of details. They are often very long but can reveal useful information about the file (i.e. a file might include 'eng subs' in its name).

- Check the size of the file. After a while you get a feel for the size each file should be. For example, a 2hr film in the .avi format is about 700MB (written as 700 million in the window as the default is set to bytes) and a normal 3 min MP3 is usually about 3MB.
- To check that all your downloads are in order, click the Downloads tab next to the search one. It will show you how much has been downloaded of each file and give an ETA. These aren't really that useful, but strangely addictive to watch.

Playing your downloads

BearShare does have a built-in player but it's awful – it's only worth using to check that you've downloaded the right file before opening another program. It's far better to open with your preferred media player.

- You can either do this by opening the file from where it's saved by browsing or setting up BearShare to do that for you. Click Setup> Options then check the box saying 'Never in the Preview/Play' in theatre part.

BearShare – Playback options

- Then all you have to do is click on the Library tab then double click on a file and it will open with its default player.

LimeWire

Like Bearshare, LimeWire is hugely popular with loads of files, but it's also riddled with spyware. To get it go to www.LimeWire.com, click the link, choosing Get Basic. When you've downloaded the software, install as normal.

- Run LimeWire from the start menu. The first time that it runs, a wizard will open helping you set up your Kazaa.
- It will first ask you what language you would like. Click Next.
- Next it will welcome you to the wizard. Click Next.
- It will then ask you where you would like your files to be downloaded. This file will be shared. Click Browse to change from the default. Click Next.
- Then you will be asked to select your internet connection. Do so and click Next.
- Next you'll be asked if you want to start LimeWire automatically when you start your computer. Choose and click Next.
- The next window will warn you that it's about to access the internet and your firewalls will ask you if you want to allow LimeWire to access the net. Click Next and give LimeWire permission.
- Once you've allowed LimeWire, click Next.
- Then you will be asked if you want LimeWire to scan your drive for media files. Select No and click Next.
- Now you've finished the set up. Click Finish.

Searching for files

To search for a file, enter the keywords into the Search box on the left of the window.

LimeWire – Search results

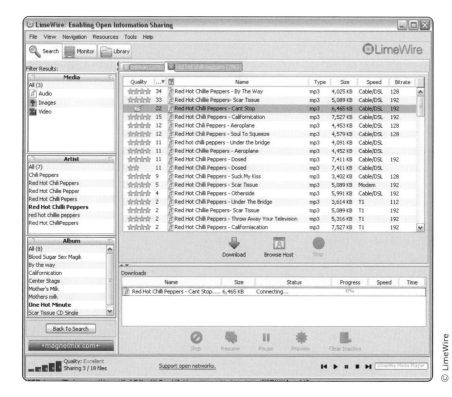

- The results will be shown in the main window. The hits will have various columns of data shown with them. The fastest to download will most likely be the one with the highest number of sources.
- Also check the file size to make sure that it is about the right size for the kind of file you're looking for. A lot of information can be gained from the file name as well so read it carefully. To download one, double click it.
- To monitor your downloads, check the smaller window below.
- To monitor the uploads from your shared files, click the Monitor tab at the top of the window.

In this chapter we looked at file sharing networks – the black art of internet downloading. If you've got this far without skipping too much, bravo. Consider yourself a black belt in file downloading. But there's one chapter left to read. It would be easy to skip over it – but that would be a mistake. So, take a deep breath, and let's plunge into the surreal world of digital copyright.

7

Copyright And The Law

Unless you've been living on Mars you might have noticed by now that the downloading of copyright material – mainly music and films – has aroused huge global legal controversy. The US Supreme Court, the US Senate and the top commercial judge in Australia have all got in on the act and weighed in with new laws and judgments.

The core of the debate is simple. Many people who hold copyright on what's called intellectual property or IP, such as musicians and film-makers, believe their ownership is being threatened by the new digital technologies outlined in this book. And although many musicians think that's not necessarily such a big problem (see the section below on what musicians actually think) it's hard not to accept that the copying of digital content has becoming insanely easy.

Despite the best efforts of software designers working on Digital Rights Management (DRM) systems, you can copy an album of music in under five minutes and a film in under an hour. And if that wasn't bad enough, new broadband technologies and file sharing networks, like BitTorrent, allow you to send that copied file very quickly across the planet.

Luckily for most musicians and film-makers they have powerful lobby groups, especially those based in the US, that have taken up the cause. The RIAA represents the main US music companies, while the MPAA battles on behalf of Hollywood, and both have considerable political and financial clout. In fact, their lobbying and funding grip on many US legislators is such that the US Senate nearly passed one act that would have put copyright infringers in jail for up to five years, even if they had only backed-up one single album and then only kept the one copy on their computer (a situation that describes many millions of average users in the UK).

And the legal bottom line is that these hugely powerful, and vocal, vested interests have a legal point. In nearly every country in the Western hemisphere the simple act of copying and distributing copyrighted material is in fact illegal. But even though many of the practices described in this book could be technically illegal, the reality is that the vast majority of us will never be hassled by any lawyer or threatened by a legal letter.

In this chapter we'll attempt to try and answer a number of commonly asked questions about what's right, what may be morally wrong, and what's definitely going to get you in trouble.

Let the questions begin.

Questions about copyright

Question 1: What's so wrong with using other peoples' copyrighted material?

It's only fair to say at this point that there are a number of perfectly respectable moral arguments that might dissuade you from copying music and film content from the internet. Perhaps the best summary of these was contained in a recent position by the British music industry federation, the BPI.

Here's what the BPI chairman Peter Jamieson said:

> "The unauthorised distribution of music over the internet is against the law. It infringes the legal rights of artists and record companies. And it's bad for music. The British record industry, which is responsible for the lion's share of the UK's investment in new artists – in excess of £150 million per year – cannot possibly hope to continue investing in new music if nobody pays for it. That's a fundamental financial fact which even the internet cannot change. Whatever your views on unauthorised filesharing, this is not good news for the wider music industry that currently employs around 126,000 people in the UK. Nor is it good for the thousands of talented British artists who hope that one day they'll make a living out of their work...The record industry simply must act."

There's a number of different arguments advanced here but in summary the BPIs perfectly reasonable case is that copying and file sharing of music and film:

- ...undermines the industry's investment in new artists

    The industry loves to quote a study in 2004 that showed music downloaders' spend on singles and albums declined by 59% and 32% respectively and their overall spending on music declined by one third; among heavy downloaders this figure increased to almost half. This, they claim, is impressive evidence that prolific users of BitTorrent actually end up spending less on music as a consequence of copying other peoples' music tracks. Another study, by Forrester Research also concluded that 40% of frequent downloaders buy less music than before they began downloading.

- *It's an unreliable and unsafe way to get music*

    Go to the big file sharing networks and you'll constantly run the gauntlet of viruses, spyware, poor original recordings and fake files. It's hard to disagree with this, but guess who's responsible for most of the fakes on the internet – step forward the music industry and their use of firms like Macrovision that use a technology called Hawkeye which floods file sharing networks with fake search results and bogus files on behalf of the music industry!

- *UK consumers DO have an alternative to illegal filesharing*

    They're called Apple, MSN Music and Wippit.

- *Its against the law*

    They're right.

Question 2: Surely it's legal to copy music from my CD onto my computer?

In the UK we have one of the strictest legal frameworks for protecting copyright on the planet. The bottom line is very simple: making a copy of a music CD or a film DVD in any shape or form is illegal.

And here's the really annoying bit: even if you bought that CD, and even if it's for your own use and you're only copying it for use on your MP3 player, it's still illegal. It's even illegal for you to copy it to tape to listen to in the car. It's all clearly an infringement of copyright.

Question 3: Hang on, but I thought it's OK to back-up your stuff if it's only for your own personal use?

Many people like to think that there's something called 'fair use' that applies to personal back-ups of their own music CDs and film DVDs. The problem is that they're wrong (in most nations at least), and they're especially wrong if they happen to live in the UK. In the UK, the law – we'll look at the actual Act in detail below – does allow certain very limited exceptions.

1. You are allowed to make back-up copies of computer programs provided they are for the original owner's personal use. Copyright law in both the UK and the US allows for a single archival copy of any given commercial software to be made, for use only if the original fails. Making any additional copies for personal use is a violation.

2. DVD movies, however, as with CDs, can't generally be reproduced in any way. But there are some exceptions here as well. Small extracts of a CD or DVD can be copied for academic use, for review or for news reporting. That doesn't mean you can get away with a full copy of the film, but it does allow you to copy extracts for a public lecture on why horror films are bad for society, for instance.
3. You can also tape television and radio shows to watch or listen to at a later time.

Question 4: All this talk of laws has got me worried – what does the law in the UK actually say?

If you have some spare time on your hands and you fancy examining the Copyright, Designs and Patents Act 1988 in all its lovely detail go online to – www.hmso.gov.uk/acts/acts1988/Ukpga_19880048_en_1.htm

The key bits of the Act are in sections 17 and 18 and they state:

Section 17 – Infringement of copyright by copying

(1) The copying of the work is an act restricted by the copyright in every description of copyright work; and references in this Part to copying and copies shall be construed as follows.

(2) Copying in relation to a literary, dramatic, musical or artistic work means reproducing the work in any material form. This includes storing the work in any medium by electronic means.

(3) In relation to an artistic work copying includes the making of a copy in three dimensions of a two-dimensional work and the making of a copy in two dimensions of a three-dimensional work.

(4) Copying in relation to a film, television broadcast or cable programme includes making a photograph of the whole or any substantial part of any image forming part of the film, broadcast or cable programme.

Section 18 – Infringement by issue of copies to the public

(1) The issue to the public of copies of the work is an act restricted by the copyright in every description of copyright work.

> *(2) References in this Part to the issue to the public of copies of a work are to the act of putting into circulation copies not previously put into circulation, in the United Kingdom or elsewhere, and not to–*
>
> *(a) any subsequent distribution, sale, hiring or loan of those copies, or*
>
> *(b) any subsequent importation of those copies into the United Kingdom;*
>
> *except that in relation to sound recordings, films and computer programs the restricted act of issuing copies to the public includes any rental of copies to the public.*

As we've already said, there are some exceptions to these clauses, that allow limited copying. These are in section 29 and they allow something called 'fair dealing' in the following circumstances.

Section 29 – Research and private study

> *(1) Fair dealing with a literary, dramatic, musical or artistic work for the purposes of research or private study does not infringe any copyright in the work or, in the case of a published edition, in the typographical arrangement.*
>
> *(2) Fair dealing with the typographical arrangement of a published edition for the purposes mentioned in subsection (1) does not infringe any copyright in the arrangement.*
>
> *(3) Copying by a person other than the researcher or student himself is not fair dealing if–*
>
> *(a) in the case of a librarian, or a person acting on behalf of a librarian, he does anything which regulations under section 40 would not permit to be done under section 38 or 39 (articles or parts of published works: restriction on multiple copies of same material), or*
>
> *(b) in any other case, the person doing the copying knows or has reason to believe that it will result in copies of substantially the same material being provided to more than one person at substantially the same time and for substantially the same purpose.*

You might also be interested to know how long these copyright protections last for:

- **Sound recordings** – 50 years from first recording or publication.
- **Broadcasts and cable programmes** – 50 years from date of first transmission.
- **Films and motion pictures** – 70 years from the death of the last survivor of the principle parties involved in bringing the production into being; the principle director(s), screen writers, etc.

Question 5: Is it illegal to download music from services like iTunes and Napster?

Here's some good news. You're in the clear if you use these services, as a limited form of copying is legal – there's nothing in the law that stops you from using a legal download service such as iTunes or Napster as they specifically license you to use their copyrighted material.

But, and this is a big but, you may find that that license is very restrictive and may have Digital Rights Management (DRM) built into it. The restrictions in that license have nothing to do with the law as such, but are determined by the commercial practices of the service operator. Some companies, like Wippit, have a remarkably tolerant attitude to how you use their music downloads. With this kind of service you can transfer the MP3 track to your portable music player as many times as you want. You can even make your own copies for your own personal use and burn them onto a back-up CD.

But many bigger operators, and specifically Napster, have a very different, very much more restrictive attitude. They use a fierce form of DRM that is built into the software that controls the downloading of content and subsequent use of the music on your computer. DRM is in fact a catch-all industry term to describe a number of technical methods used to control or restrict the use of digital media content on electronic devices. It can come in a number of different shapes and sizes. With expensive video editing software systems like Avid, for instance, you're required to have a USB dongle that must be attached to the computer using the software, while some less expensive software systems require you to activate your software over the internet.

Movies also have their own, very varied, levels of DRM. Thanks to the different worldwide release dates for DVDs, Region Encoding is one copy protection mechanism you may be familiar with already. Another form of DRM is Content Scrambling System, or CSS, which is designed to stop you ripping data to your hard drive. Outfits like Napster and Microsoft use another form of DRM for online based services. These control the way you copy, convert or transfer the music and they work by controlling specific activities like converting your music tracks into WAV files so they can then be burnt onto a CD. For that privilege you'll have to pay extra!

Question 6: I'm worried by the rumours that there are technologies that will crash my computer if I try to rip a CD or DVD?

Some of you may like Celine Dion (although I can't say I do) and some of you may even have bought her 2002 release called *A New Day Has Come* (not her best).

In your excitement to have a long-term copy of her work, you may have decided to technically break the law and back it up onto your PC. If you had done so, you might then have encountered a particularly nasty form of DRM and you might also have encountered the blue screen of death (you know it by now: the computer freezes up, stops working and crashes...you curse...you have to run a reboot etc. etc.).

Dion's music had been impregnated (so to speak) by a particularly zealous piece of DRM. The good news is that extreme DRM like this is still fairly rare, but the bad news is that the music industry is embracing DRM with increasing fervour.

But don't despair!

Despite all the grand sounding technology, many hackers can break DRM systems surprisingly quickly. Computer science graduate John Halderman, for instance, demonstrated how to override one such system, SunnComm's MediaMax CD3 protection software, simply by pressing the Shift key. Virtually every music CD on the market – including Celine Dion's album – can now be broken using mass market ripping systems, and video-based DRM systems are also under assault from video hackers.

Halderman's clever but simple work around, one of many, underlines an important reality check: despite all the hype, most DRM is actually fairly harmless and can usually be cracked. It's also important to understand that not everyone even within the music industry is terrifically happy with this kind of technology. The giant electronics and music company Philips, for instance, which pioneered the CD format with Sony, has hit out at the copy protection methods, and threatened to remove the compact disc logo from such CDs.

> **Warning**: A warning on uncrackable DRM systems – the DRM used in online music services is much more effective and, to date at least, has not been broken or hacked. You can't, for instance, buy DRM-based music (in the AAC format) from Apple's iTunes store and try and convert it to non-DRM MP3. If you try, it may not crash your computer, but it certainly won't copy across and be playable in any device on planet earth!

Question 7: What happens if I distribute my copyrighted material (music or film) online using file sharing networks – will I get into trouble?

Imagine that you've ripped some music from a CD and you decide to upload that music to a file sharing network using something like BitTorrent. Is that illegal? The answer by now should be blindingly obvious: yes, it's illegal! As the law stands, all distribution of music and film that belongs to another copyright owner is strictly illegal. More importantly, it's also *very* likely to get you in trouble.

Here in the UK, the legal lead has been taken by an organisation called the BPI, which has specifically targeted uploaders. It started its campaign a few years back with some sensible emails to prolific abusers. The reasonable sounding messages pointed out that uploading other people's copyrighted material was against the law and generally a bad idea.

Needless to say most dedicated file sharers simply ignored the message and carried on regardless. A few months later the BPI started to get tough! They sent out a flood of threatening legal letters to a small hardcore of what they called 'serial uploaders', who offer hundreds if not thousands of music files over the internet. The logic behind this is obvious. The BPI says 15% of file sharers are responsible for 75% of all songs swapped illegally, so if you stop the big uploaders who serially place music on networks like BitTorrent you'll stop most of the illegal material.

And there are certainly some very serious serial uploaders around. The largest case this side of the Atlantic so far was in France where one uploader had 56,000 tracks (equivalent to more than 5,000 CD albums), in his online music library. But you don't have to upload and make available thousands of tracks to receive a nasty legal letter. People who have uploaded even a few hundred tracks are in the spotlight in the latest legal cases.

The most recent legal campaign in the UK has focused on a bunch of music fans forced to pay up to £4,500 each and agree to accept injunctions against them illegally uploading music in the future. The 23 settlements came after a series of letters were sent out in October 2004 threatening future action. The offenders included seventeen men and six women, aged between 22 and 58, a student, the director of an IT company and a local councillor, with the average settlement of £2,000.

Here's what the BPIs lawyer had to say:

> "We have no desire to drag people through the courts. So we have attempted to reach fair settlements where we can. We hope people will now begin to get the message that the best way to avoid the risk of legal action and paying substantial compensation is to stop illegal file sharing and to buy music online, safely and legally, instead."

So, that means settle up now and stop uploading or face a big court case.

Question 8: I hear that in the US the threats against uploaders have been even more severe – I don't have to worry about that if I live in the UK do I?

The focus of most big, high-profile legal actions has been in the US, home of not only the global music and film industry, but also of most file sharing network users and most high-profile electronic and technology companies. In the US, over 10,000 people have been threatened with legal action, and most of them have had to settle up and pay tens of thousands of dollars in costs.

The US has also taken the lead in a number of other battles against infringement of copyright. The big music companies managed to shut down the first really big file sharing network, Napster, in 2000. After a lengthy legal battle, the music industry managed to prove that the network itself – it was based on a centralised server that held details of all the music of all its users – was knowingly involved in the infringement of copyright.

Napster was only the first of a long list of targets carefully chosen by the music and film industry as part of a strategy to completely change the legal footing of copyright protection in the US. Their biggest success has been something called the *Digital Millennium Copyright Act* (or DMCA). The DMCA, introduced in 1998, was designed to update American copyright laws to deal with the new issues posed by the internet and the proliferation of powerful multimedia home computers. The DMCA currently considers the bypassing of a copy-protection scheme put in place by a copyright owner to be a criminal act; even if the person doing the bypassing purchased the copy-protected material legitimately. To put it simply, you cannot legally duplicate copy-protected material, even if you own it.

This section of the DMCA (chapter 12) has been massively controversial, since not only can it remove the right of consumers to duplicate their purchases, it also makes any tools that can be used to duplicate a copy-protected product in violation of the copyright of that product, and thus the makers of these tools become vulnerable to civil law suits. So if you're a company selling software that breaks copyright protection of movies for instance, you're in big trouble. Which is precisely what happened to a company called 321 Studios.

321 Studios

321 Studios was one of a number of companies that supplied software that enables people to make perfect digital copies of films, but the movie studios argued that this permits illegal copies to be made and circulated, infringing their copyright. In response 321 said the software allows people to make back-up copies of DVDs they have already paid for, and it fought its way through the US court system and lost. In its mission statement, 321 Studios, said that:

> "Consumers who copy DVDs for their personal use are exercising their right of 'fair use' – an exception to copyright law that has been upheld by the US Supreme Court in order to avoid an irreconcilable conflict between copyright law and the First Amendment's guarantee of free speech."

Here's how the software in question – DVD X Copy – worked. It allowed a DVD movie disc to be copied to a blank DVD, even though the compressed MPEG code on the disc is usually digitally encrypted using something called a Content Scrambling System (or CSS). CSS makes conventional copies unplayable. But DVD X Copy doesn't hack the CSS protection, which would explicitly break copyright laws. Instead it uses a conventional software-based DVD player system to unscramble the MPEG code for legitimate playback, and then sends it to a temporary cache file on the PC's hard drive. From there DVD X Copy copies the unscrambled MPEG code onto a blank DVD using a conventional DVD burner.

But DVD X Copy had one other big selling point. Most film DVDs are pressed onto dual layer discs with about 9GB capacity, while blank DVDs have only 4.7GB capacity. So to get a film onto a single over-the-counter blank DVD you need two discs for one DVD9 (as it's called) movie. DVD X Copy, along with other programs like DVD Shrink (which is much better and free!) get around this by adding extra compression into the way the program deals with this MPEG code so that the entire movie can be burned onto a single blank DVD.

Unfortunately, this cleverness also opens up this kind of software to an obvious charge: if you buy the software you are obviously going to copy something that has had clever copyright control software built into it to stop you. Under the DMCA Act this bypassing of copyright protection is itself illegal. So 321 lost its action and had to withdraw certain versions of its software and additionally put warnings on the software pages telling its customers not to use the software with copyrighted material. Not that that got 321 Studios very far – the huge cost of the legal action eventually caused it to go bust.

Targeting non-US nationals

The American music and film industry have also started to use American laws to fight individuals living abroad.

Normally what the Americans choose to do in their back garden is strictly their business, but the case of two Brits – Kevin Reid and Ian Hawthorne – shows that we're not necessarily safe even in the UK.

Reid and Hawthorne ran a website called bds-palace.co.uk, which linked copyrighted material to the global BitTorrent file sharing network. If they didn't co-operate with a US-based action they were told that they might have to pay $150m in damages. Settle quietly and they would only have to pay $7m!

The demands, from a Washington DC law firm called Jenner & Block, called for Reid and Hawthorne to hand over the identities of the alleged copyright infringers using their website. Needless to say Reid and Hawthorne have refused to pony up the money and settle quietly. And unless they fancy visiting the US any time soon it's a safe bet that the legal action won't get anywhere in the UK.

Anyway, Reid and Hawthorne have always said that what they were doing was legal in the UK. They point out that BitTorrent does have its legal uses and bds-palace.co.uk links to legal content hosted by that network. Where they have been notified of links pointing to illegal content, they have removed those links, they claim.

The legal bottom line

Hollywood's lawyers would have to fight their legal actions in a UK court. They'd have to convince our legal system that the evidence from the US should apply here too, or challenge Reid and Hawthorne under UK copyright law. It's unlikely to happen but the threat of legal action in the US still hangs over another UK citizen, Alexander Hanff, who has also been threatened by the music industry body, the MPAA, over his website DVDR-Core.

Question 9: Is every country quite so fierce in its attacks on people who illegally use and distribute copyrighted material?

The tough American and British actions against copyright infringers shouldn't be viewed in isolation from the global assault now underway. The record industry is currently targeting nearly 1,000 people in a new wave of lawsuits against alleged 'illegal song-swappers' in actions in 11 countries in Europe and Asia. Following its first year of legal actions in Europe, which resulted in 248 people paying fines or facing sanctions, the International Federation of the Phonographic Industries (IFPI) has said it's bringing lawsuits to four new European countries, specifically the Netherlands, Finland, Ireland and Iceland. Japan is also getting in on the act as it becomes the first Asian country to take legal action against people who use file sharing services to download copyright material.

These global industry lobbies are carrying out these campaigns because past evidence suggests they work. A huge blitz of actions in Germany a few years back has already resulted in illegal file sharing falling by one third. "The number of music files downloaded there fell to 382m files in 2004, compared to 602m the previous year" says the IFPI.

But, in true contrarian, Gallic style, some French judges are starting to openly question these aggressive campaigns. Here's what one leading French magistrate told the technology magazine, *Wired*.

> *"We are in the process of creating a cultural rupture between a younger generation that uses the technologies that companies and societies have made available, such as the iPod, file download software, peer-to-peer networks...."*

The holder of such unorthodox views (to the music and film industry at least) is a certain Judge Dominique Barella who also happens to be the president of the all important French magistrates union. He's leading a fight back amongst French magistrates who believe they have much better things to do with their time than threaten iPod users with jail. "It's like condemning people for driving too fast after selling them cars that go 250kph", says Barella.

Needless to say the French music and film industry didn't react warmly to this declaration. In a letter to the French government the industry announced that it was:

> *"Surprised and shocked that the president of the magistrates union, given the level of influence he has on his (judicial) colleagues, can publish in the press a call to not criminally sanction criminal acts, which contradicts the intentions of government bodies."*

According to *Wired*, the letter also ominously thanked the minister in advance for "taking actions that he deems appropriate."

Question 10: Now you've got me really worried. I've also heard that these file sharing networks like BitTorrent, Grokster and Kazaa are themselves illegal?

A great many internet users steadfastly avoid networks like Kazaa and BitTorrent because they've heard that the networks themselves are illegal, even though some of the content on them is actually not under any copyright protection at all. Are they right? Unfortunately there's no simple answer to this important question.

The key legal battle is in the US and centres on a high-profile campaign by 28 of the world's largest entertainment companies against the makers of Morpheus, Grokster and Kazaa.

This huge legal battle has been fought all the way to the US Supreme Court, where the music and film industry companies won a notable battle in 2005. By a significant majority the top US Judges ruled that Grokster and its allies were guilty of infringing other people's copyright and that their software – peer-to-peer software used to run file sharing networks – should be amended or withdrawn.

It's worth going over some of the details of this case because it sheds light on a profoundly important debate that's been raging for decades between electronics companies and media outfits in America. The core issue is whether it's right to sell a piece of equipment or software if it can then be used to copy other people's music or film (their intellectual property), even though the original supplier of the equipment/software has no knowledge or even control over the final consumer.

The precedent of VHS recorders

Many campaigners maintain that what's at stake here is the very essence of technological innovation itself. If an equipment supplier who makes a VHS player, for instance, has to make sure the device won't ever be able to be used to copy somebody else's copyright material, won't that stifle innovation?

Many years back the US Supreme Court sided with this point of view when faced with an action by Universal against Sony, the makers of the Betamax video machine. Universal argued that these new fangled video machines were clearly designed to copy films broadcast on TV and thus they should have some form of copyright protection built into them. Sony, not unreasonably, argued that they couldn't police what the buyers of their video machines did with them and that many of the uses of these machines were perfectly legal. What's more, by forcing them to insert some kind of technology that stopped copying, their innovative technology would be stopped dead in its tracks. Sony won this battle (but lost the war of video formats as VHS eventually dominated the world) and that victory has, until recently at least, stood the test of time. In essence, new technologies and products are legal and cannot be held liable for consumers infringement of copyright so long as the device or software is capable of substantial non-infringing uses.

Over the last few decades the music and film industry has fought long and hard to overturn this decision. The RIAA even took an early pioneer of MP3 players – Diamond Multimedia – to court for devising a technology that was clearly designed to use copyrighted music. Again the RIAA were not successful but had the decision gone against Diamond Multimedia and the technology it pioneered, Apple's hugely successful iPod and the legitimate online download service that it has spawned would surely not exist today.

Needless to say the music and film industry see technical innovations, like video recording, MP3 machines and file sharing, in a very different light. They think it's very clear why consumers are buying these innovative products – to copy other people's music and film.

MGM vs Grokster

More importantly, the legal advocates for the entertainment industry have long argued that technological suppliers have:

> "turned a blind eye to the infringement of their users, and that by turning a blind eye to detectable acts of infringement for the sake of profit gives rise to liability."

That summary is from a legal argument put to the Supreme Court in the case known as MGM vs Grokster – a key action that has turned the legal tide in favour of the entertainment companies.

Grokster lost this particular case for a number of complex reasons. First there was copious evidence from Grokster and its allies' (notably a network called Streamcast) own internal papers that they were targeting ex-users of the now defunct Napster service. Napster had been originally shut down because it was patently clear that the service was promoting its business to users who wanted to get illegal content. One marketing campaign proudly implored users to join Napster and 'Get Hundreds of hours of music for next to nothing'! The courts held that Napster should clearly be liable for its users' infringement of copyright, and thus, if Grokster and its allies were trying to poach these customers, they too would be liable.

Also, neither Grokster nor Streamcast have tried to develop any filters or clever little tools to control the copyright infringement that was clearly going on using their software. At one point it was estimated that 90% of all content on Grokster was illegal. If they'd at least tried to advertise some filter that purported to stop all these illegal music and film files, then they could at least claim they'd attempted to do something; but in reality they hadn't bothered, so they were guilty!

The last nail in the coffin was perhaps the most obvious. It was always pretty clear to all and sundry how Grokster and Streamcast made their money – from advertising – and more specifically from advertising that was paid for on the basis of the number of eyeballs looking at the website. But what drives more and more people to use the networks and thus look at the advertising? Illegal content of course! People use networks like Grokster because there's an enormous quantity of useful, illegal stuff on them, and in reality, the Supreme Court argued, the management of both Grokster and Streamcast knew that too and that's what determined the rates they charged to their advertisers.

What this landmark ruling in effect created was a new theory of copyright liability that measures whether manufacturers create their products, services or software with the 'intent of inducing consumers to infringe'. It's also important to underline what this judgment didn't say. It didn't say that file sharing networks, per se, were illegal. It merely concluded that the owners of Grokster and Streamcast knew they were being used to distribute illegal content and did nothing to stop it.

The name of the game now – post-ruling – is this: if a file sharing service clearly declares that copyright material *should not* be used on the network and that this kind of content is illegal and then devises some (fairly pointless) filters that allow you to bypass the illegal content, that service can be legal.

In fact, the Supreme Court specifically stated that they did not regard their judgment as a barrier to people devising technologies that might involve copyright infringement – their only concern is that the innovators don't knowingly market a service or product aimed at copyright infringement.

The approach in Australia

Courts in other key countries have taken a very different, more stringent view.

In Australia, for instance, home of the Kazaa file sharing network, a Federal Court judge has clearly sided with the entertainment industry and ruled that all file sharing networks are in effect illegal, unleashing legal claims that could total at least $760 million.

In Justice Murray R. Wilcox's ruling, the file sharing networks, and Kazaa in particular, were found guilty of enticing Australians to make and distribute unauthorised sound recordings without license, as well as collectively "entering into a common design" that enabled unauthorised trafficking. Kazaa either had to shut down or design a search system that excluded copyrighted works, and broadcast messages to current users compelling them to upgrade to this rights-managed version.

But the judgment also clearly pointed that "the internet is here to stay", that P2P is a force unto itself, and that rather than fighting the natural course of evolution, the recording industry should adopt stronger safeguards and digital rights management procedures to protect themselves. If CDs couldn't be ripped the court action wouldn't have happened!

Conclusion

So, returning to our original question of whether these networks are, by design, illegal, the answer is: probably maybe! If the file sharing network was specifically designed to share illegal content and the developers and owners of the network knew that, the answer is, in the US and Australia at least, yes, it's illegal.

But file sharing networks and their technologies are not, in pure legal terms, illegal and if the operators of these networks issue loud warnings about illegal content and try to devise content filters, then that network would be legal.

And lastly, there's the small matter of UK jurisdiction. To date, no-one has actually fought in the British courts to close down file sharing sites and thus there is no major UK court case that suggests file sharing is illegal.

Question 11: Am I going to face legal action if I am only an occasional downloader, accessing just a few music tracks and the odd film?

Let's pause for one moment and reflect on a smart article by the journalist Tony Smith of the Register, an online technology newspaper. Smith has actually bothered to do his homework on the law behind the internet and file sharing and his analysis of the legal campaign against the big serial uploaders has some interesting implications for the occasional downloader. His key point is that the big legal actions against users who upload thousands of tracks is doomed to fail. As Smith says:

> "There are going to be limits to the effectiveness of this route. The actions will make an impression on the general public, but the message that gets across is more likely to be that it's really not smart to be sharing large numbers of files, rather than 'It Is Stealing and It Is WRONG'. So they share less files, and as the memory of the publicity subsides, they slowly share more files. So they have to be reminded. And reminded again, and again. If the music industry only goes for the big sharers, then it really can't hope to do much more than damp it down, and the more it does it, the more obvious it will become that modest levels of sharing are, apparently, safe. Which means that if they don't do something broader, it's even arguable that they'll be making most people more confident that they're safe from lawsuits. This presents a nasty choice to the BPI – widen the actions, or view the current campaign as a limited, short-term measure and think of something more sustainable."

I've quoted this article at length because it raises an important point. To paraphrase Smith, the legal action against uploaders is only ever going to be a campaign against a small number of people. If they do decide to target the hundreds of thousands of occasional downloaders in the UK, they'll be fighting in the courts for decades. In essence, it's all a 'scare' campaign that will probably end up stopping very little.

Question 13: What happens if it's my children doing the illegal file sharing but I receive a letter threatening legal action?

The BPI says it can't differentiate between children and adults in its legal actions because it does not actually know their identities. Typically the BPI will track the big file sharing networks and try to spot prolific distributors of copyrighted content. They'll quickly be able to track most users back to their internet service provider – the ISP then receives a request from the BPI's legal team demanding personal information. Some ISPs have resisted these orders and in other countries, such as Holland, ISPs have fought their way through the courts to fight the music and film industry legal actions. But in the UK most ISPs will willingly oblige. Your details – not your childrens – get handed over to the BPI's lawyers and within a few weeks a letter arrives in the post. And as the adult in the house, who's signed up with the internet service provider, you are responsible and you will probably have to pay the legal costs as many parents in the UK have already discovered!

Question 14: Has anyone asked the musicians what they think about these file sharing networks and the threat of legal actions?

The music industry loves to make out that it leads a united front in the battle against the despicable music pirates. The clearly stated assumption is that the industry is fighting these legal actions globally to defend the poor musician. And certainly whenever this issue makes it into the media in the US a friendly music act like Metallica are wheeled out to complain about their livelihood being threatened. But what do the vast bulk of professional musicians actually think about all this digital music and file sharing?

Luckily the US-based The Pew Internet & American Life Project has conducted a survey among musicians. Here's what they (surprisingly) said:

- 66% of musicians surveyed said that the internet is 'very important' in helping them to create and distribute their music. Additionally, 90% of all respondents use the internet to seek out inspiration.
- A massive 66% of musicians' view file sharing as a minor threat or no threat at all.
- Musicians don't particularly like file sharing networks that distribute their music online without any payment for copyright. Two-thirds of all artists surveyed said that the people who run the file sharing services should be held responsible for the piracy, while 37% of musicians believe that both those running the file-sharing programs and the individuals sharing the files should be held jointly accountable.
- 60% of all musicians surveyed said the US music industry's legal actions will not benefit them in any way.
- Selling material without the creator's permission is broadly regarded as wrong, while copying material for private use is seen as acceptable.

It's also worth making one last point, namely that a growing number of musicians do use networks like BitTorrent and Kazaa to distribute their music to a new audience – look at the recent success in the UK of the Arctic Monkeys. If you're an unknown act, file sharing networks can be a great way of building a fan base who can then be encouraged to buy the album via download from the band's own website. The Metallica's of the world may have the enormous advantage of a high-profile music label to publicise them, but an increasing number of musicians think file sharing networks are their best allies.

How can a serial uploader avoid attention?

1. My best advice is to get off the traditional file sharing networks based on the FastTrack system (Kazaa/Grokster) and head off to BitTorrent – the big American anti-piracy groups like the RIAA are concentrating most of their efforts against FastTrack. If you can avoid this network, the chances of getting caught are diminished substantially.

2. Make sure no-one can see your shared directory. If John Doe can look in your shared directory then you can bet your life the anti-piracy watchdogs will be looking as well, especially if you've got loads of films and music in there. It's against the spirit of file sharing, but get real…

3. Share fewer files. If you're a P2P true believer, the idea of cutting down on what you share is probably blasphemous, but as with hiding your shared directory, it makes sense. Have thousands of music tracks in your directory rather draws attention to you. According to the file sharing news community Slyke, the majority of RIAA targets have 800 or more shared files. If you feel you do want to share your music and want to avoid trouble, keep the number of files below 50.

4. A lot of smart file sharers are heading off to the communal networks based on the IRC channels or sharing within tightly controlled trusted networks based on friends. If you share with who you know you'll probably be safe most of the time.

5. Remove all potentially misleading file names that might be connected with the name of a leading US-based artist.

6. According to one organisation, the Electronic Frontier Foundation, the RIAA also appears to be targeting its legal actions at users who allow their computers to be supernodes on the FastTrack P2P System (used, for instance, by Kazaa and Morpheus). They say to reduce this you should make sure your computer is not being used as a supernode.

7. Some file sharers are using a bit of software called PeerGuardian. This blocks bad IP addresses from your computer – these IP addresses belong to the anti-piracy investigators. It does this by using a publicly available databank of IP addresses from an outfit called Blocklist.org. This is a clever idea (and one I use), although I presume the big boys – the RIAA and MPAA – know their IP addresses are in public circulation and simply switch server and address or use a technology called dynamic IP address.

Sounds like the case for copying and downloading is legally and morally hopeless?

There is in fact a strong moral case for fair use of copyright material you've purchased. The law in the UK may not accept that, but other legal jurisdictions do recognise that some uses of copyright material are acceptable. The problem is that the additional moral case for then sharing that content is dubious to say the least.

The moral case for fair use

With most things we buy, we acquire the right to use them in any way we want within certain limits. If we buy a car we buy it because we want to do anything we want with it, with the exception of maybe using it to kill someone with it at great speed. So we accept that property rights infer on us exclusive control within certain limits designed to protect the public.

When you buy media content – films or music – you only buy a license to that property for your own private use, not to distribute around the internet. In effect, you're not buying the music, per se, but your copy of it. The entertainment industry maintains that if you copy this material in any way, that infringes the ultimate owner's rights.

But let's go back to that example of the car. We accept that any limits placed on us in the use of our newly acquired thing (a car) are to safeguard other users from significant injury. But the limits placed on us by the music and film industry are not to protect against injury or harm – in reality the restrictions placed on our use is simply to defend the intellectual property of the owner.

In and of itself that's no bad thing. If I paint a great work of art after decades of hard graft, the odd bout of insanity and copious quantities of drugs, I should have the right to benefit from my newly painted property. I'd be especially annoyed if some faker came along, exactly copied my work of art and then sold eight thousand of them over the internet. Defence of property rights, especially against people who wish to copy my work and then distribute it widely and freely, is perfectly reasonable and fair.

But what happens if I buy that work of art, as a rich collector, and then choose to make a facsimile copy of the work, which is in turn stored in a safe place in case my house is burnt down. Surely this form of copying, to protect my investment in this work of art is reasonable and fair (if probably a bit sad). I'm not attempting to profit from the artist just to protect my investment.

In a sense the copying of a CD or DVD for personal back-up use is the exact equivalent. I have paid money for my own copy of a work of art, and this CD and DVD is valuable to me and I want to back-up my investment, especially as CD and DVDs are hardly indestructible. One small scratch on a DVD could wipe out a £20 purchase!

What is so wrong with backing up that investment and then using it for my own purposes? Many music industry license agreements already tacitly accept this argument by granting legitimate users the right to store the digital content supplied by the network on whatever storage media they want – to, in effect, safeguard their property (Wippit, eMusic, et al). Equally, the software industry also accepts that users may want to make back-up copies of essential software. What's so different about backing up a hard copy, physical CD or DVD with film or music on it that you've also paid for?

And let's be realistic here, the entertainment industry already tacitly accepts this fair use point. They may proclaim that any copying is illegal, but to date no-one has been prosecuted in the UK for making personal back-ups, solely for private use. The industry recognises that millions of us do it, and it's not worth them stopping us from doing what many of us regard as fair.

Isn't file sharing a way of making content accessible to millions cheaply?

Backing up content for personal use has a strong moral argument in its favour, but the act of then sharing that property over a public network is a much more difficult moral argument. There's no getting away from the fact that you are effectively sharing someone else's intellectual property.

Some file sharing activists say that as long as no money passes between the participants in a network, what's the problem? Even they'd accept that buying pirate DVDs on a street corner is a clear example of someone stealing property from someone else, but file sharing networks don't use money, just co-operation and trust. But the bottom line is that someone else's intellectual property, on which they financially rely on, is still being taken away and not paid for. Whether you call this theft or copyright breach is a moot point, but it's still probably not right.

Many file sharers fall back on another argument: that file sharing is fair because film and music content is just too expensive. They have a point. The high street price of CDs and DVDs is nothing short of legalised larceny, especially in the UK. Visit the US and the net high street price is just under half of what we in the UK are charged. And music prices seem to have been going up steadily for years – high street prices for top albums have recently started climbing above £15 an album, well up on the price of vinyl which is presumably a more expensive medium. Equally, the price of movie seats, especially in the big cities, has been steadily rising for the last decade, and in London it now regularly tops £10. It's difficult to argue with those critics who maintain that we, the consumer, are being fleeced by the entertainment industry.

All of this criticism sounds immensely reasonable, up until the point at which you upload some copyright content on a network and then make it available to all and sundry for free. If you don't like the prices being charged for CDs and DVDs, find somewhere cheaper to buy the product! Go online and buy the CD or DVD from a Jersey-based operator that doesn't pay VAT! Don't like cinema prices, why not wait until you can rent it on a DVD, and then watch it on your own home cinema system? The big movie studios and record companies are probably ripping us off, but that doesn't justify ripping them off. Two wrongs don't make a right. All you have to do, as an active consumer, is fight back and find the cheapest way of legitimately purchasing the CD or DVD, not copy someone else's copy.

File sharing – a force for the good

Whatever you may think of the moral basis of file sharing, it has at the very least produced a series of tangible public goods.

- File sharing has sparked a series of technological innovations, such as the Skype voice telephone calls over the internet service and, indirectly, the legal music download services that have created billions in sales and created thousands of jobs.

- File sharing networks have also forced the music industry to invest even more money into innovation and developing new acts and new ways of getting performers in front of music fans. More and more music companies make more and more of their profits from carefully managed series of live acts and performances.

- File sharing has also forced the music industry to accept that not everyone wants to shell out an awful lot of money on an album with just a few good tracks. Music consumers are becoming much choosier about their purchases and are increasingly unwilling to accept album filler material.

Conclusion

The not too distant future – what to watch out for

File sharing is a wonderful driver for change. It's forcing those lazy media giants to open up their vaults and start releasing content on increasingly fair terms. As I pointed out in our last chapter on the legal position, file sharing is a powerful force for the good, encouraging wonderful new innovations that should make your entertainment life that bit richer.

But this wonderful change is also going to throw up some new challenges for you. Here's my list of 10 things to watch out for in the near future.

1. Avoid the soon to be launched legal online film download services

As I write, Universal is rumoured to be about to start offering a service whereby you buy a DVD, and get the right to download one copy of it to your hard drive. In theory this is a great innovation. But it's also almost certainly a waste of the extra money you'll probably have to pay for the privilege, because that downloaded film file will have fierce DRM protection built into it, which restricts your ability to make a back-up or watch the data file on your home TV. The bigger point here is that movie downloads will be expensive, fenced in with astonishingly restrictive DRM and probably won't have the really interesting films you're after. Personally, I'd sit out this market for the next few years and wait until a real innovator with power, like Apple or Amazon, wades in.

2. Broadband speeds – here comes 24 Mbps!

The whole broadband ISP space is one of the fastest growing markets on planet earth and there are new entrants – and new technologies – emerging almost every week. Price is what is fundamentally powering this competition, but new technologies will only serve to intensify this battle and the next big one will be over upgrades to your broadband speed. 8Mbps is next on the agenda in the UK, followed by super ADSL technology that will kick speeds up to 24Mbps. Cable companies might even start experimenting with 100Mbps speeds. It all sounds wonderful, until you ask yourself why you need all this extra speed and capacity? In my opinion, 2Mbps for most heavy downloaders is probably more than adequate for the time being. Obviously jump on board the 8Mbps express when prices crash down to between £15 and £20 per month, but until then sit it out.

3. Watch the wireless broadband space carefully

The big mobile networks are currently rolling out data cards that fit into laptops that can run at broadband speeds for about £30 a month. But there's a catch – your usage is capped at around 250MB a month, which makes it useless for any kind of file sharing on the move except maybe for music. But the technology is evolving fast and I'd expect prices to fall sharply and smarter phones to emerge that can also surf these 3G networks. And then there's also WiMax. It's a little over-hyped at the moment, but it could potentially disrupt everything. It's essentially a Wi-Fi network that extends over many miles. This kind of broadband public access infrastructure could open up wonderful new possibilities, especially as big cities start to compete with each other to offer the best range of hotspots.

4. Good old-fashioned ethernet technology still works!

In the great rush to embrace the wireless age many PC users forget that cabling via Ethernet works brilliantly. As a technology it boasts a number of attractions, notably it's secure from wireless hackers and it's reliable over longer distances with big multimedia files. If you plan to share media rich content around your house, in different rooms, with different devices and PCs, think long and hard about an ethernet network. It will work particularly well with transferring big data files, and it's cheap and relatively easy to configure.

5. The next big thing – voice over the internet

Broadband and file sharing technologies have given new telephony technologies a huge push and the next big thing is called 'voice over the internet' or VoIP. One of the leading operators in this space is Skype, which uses a kind of file sharing technology to enable its users to make ultra-cheap or free calls anywhere around the world. VoIP has emerged as a mass market phenomena and the technology works. It's also ultra cheap; and if you sign up with operators like Vonage you can use the service even when your computer is not switched on.

6. P2P activity will not diminish

Renegade file sharing technologies will keep coming under sustained attack over the next few years and the big entertainment companies will try to convince you that P2P is on the decline – don't believe them. It's a propaganda war now and the anti-piracy police have a vested interest in predicting 'the end' – the demise of file sharing – but the file sharing technologies will carry on growing and keep finding ways of shaking off 'the enemy'. The immediate front line in this battle though will be the big indexing sites for the BitTorent network – TorrentSpy is already under full attack and I'd expect an assault any day now on the excellent IsoHunt.com.

7. Another next big Thing could be legal file sharing

This allows you and your mates to share files easily and safely on 'closed user' file sharing networks. This technology has been around for a long time, and many users of BitTorrent already use this technology, but there are a raft of new providers emerging in this space with some great products. In my experience a lot of the proprietary networks are tricky to use and the best of the bunch at the moment is something called FolderShare which you can use for free.

8. To HD or not to HD?

High definition is coming fast and already many of the main file sharing sites boast dozens of files with the letters HD somewhere in the melange of acronyms. Although downloading HD files specifically is a bit of pointless exercise. Usually most of the files have been compressed to buggery and thus all the great virtues of HD – better line resolution and crystal clear picture quality – vanish almost instantly. Nevertheless, if you see two files that are exactly the same except that one is HD, obviously download the HD file – the picture quality will be better.

9. Portable disk drives versus streaming devices versus media centres?

In this book we've already mentioned the growing assault on the lounge by PC technology. The next great battle is going to be over the device that makes it into your lounge – will it be a very easy to use portable hard drive that connects straight to your TV, a clever black box that streams content to the TV or will it be Bill Gates' great vision of a consumer electronics new world – media center PCs?

It's very early days in this enormous battle of the future, but I'd hazard a number of important guesses. First off, media center PCs that really work as promised are probably still too expensive. You need devices that really do work easily without any real meaningful internal fan sound, and at the moment that means spending £800 or more. Many of you might be tempted to go down the DIY route and build your own smart entertainment PC that uses Windows Media Center as an operating system, but beware – it's a very choosy operating system that doesn't work well with many off the shelf components that go into a PC.

Streaming devices are already battling it out in the music space and most of them work very well, but Pinnacle has recently upped the ante by releasing a media device that sits under the TV and plays video off your PC. It's a great idea and my system works brilliantly, but I'd be lying if I didn't admit that most firewalls struggle to cope with it and the networking technology that lurks behind it is a bit fiddly. Many consumers, by contrast, are tempted by the cheap hard disk based 'brick' devices where you transfer movies and music off your PC and then re-connect to the TV using a Scart connector – we mentioned them in our section on video and featured the LaCie player. The technology absolutely works and is very easy to use – and what's better it's increasingly cheap and good quality devices shouldn't cost you more than £150 at most nowadays.

My verdict? Don't even look at Media Center PCs until costs have come below £500 and the OS has been simplified, while streaming devices are fine if you are technically savvy but should otherwise be avoided. Hard drive players are the simplest and best technology on the market at the moment (Rhapsody do a particularly good one), and new devices that also incorporate wireless links and even a streaming capability are also on their way.

10. Look out for iMP from the BBC

The venerable British Broadcasting Corporation deserves all the plaudits its received over the last decade for its technological breakthroughs. Its news websites really are fantastic – especially now they've built in RSS support – and its trials in the podcasting space are superb, with its weekly *In Our Time* download (a philosophy, science and history discussion series) taking top prize for fabulous, compelling content available for free with ease. The next big thing will be its online video player called iMP (Interactive Media Player). This will allow online viewers to watch most of the BBC's output for around seven days

Conclusion

after its broadcast, probably for free in the UK at least. It should be due out sometime in 2006.

Users of a similar video on demand service, pioneered by the company HomeChoice, have already test trialled this fantastic service. And while on the subject of HomeChoice, if you are in the right area to get this service, subscribe now! This is the future now, and allows users to access nearly all the key cable channels via their internet connection and also lets viewers watch films and certain popular TV programmes on demand, whenever they want to.

BT are rumoured to be about to bring out a challenger to this service but video on demand using the internet is a thing of wonder and completely changes the relationship of the viewer to TV. As with other download technologies it allows you to watch what you want to watch, when you want to watch it. The only big downsides in my personal experience are that the picture quality isn't fab, and you can't actually record the programmes and store them on your hard drive for later use. Still, the world's not a perfect place and as it only costs around £25 a month with ordinary 2Mbps broadband thrown in plus free calls, beggars can't be choosers. Video on demand using broadband has arrived and it'll completely change the way we use TV...

Index

Index

A

ActiveX, 55, 61-63, 82, 139

Analogue, xvii, 23-24, 85, 134

Anti-piracy, 244, 247, 252, 261, 263, 268, 300, 309

Anti-virus, iii, 39, 43-46, 52-53, 68, 82

AnyDVD, 193, 211-212

Apple iTunes, iv, xvi, 93, 103, 116, 133, 145, 168

Atom films, 187-190

Audacity, iv, 123, 127-129

Audible, iv, viii, 133, 174-177

Audio Recorder Pro, iv, 123-125, 127, 129

AutoXDCC, 247

AVI, 100, 196, 237, 254, 275

B

Back-up, x, 12, 14, 82, 199, 209-210, 212, 214, 217-219, 234, 238, 283, 286, 290, 302, 307

Backing up your dvds, 187

Bandwidth, 243-245, 249, 269

BBC, iv, 120-123, 190, 310

BearShare, 246, 268, 273,-276

Bitrate, 126, 162, 219, 225, 233

BitTorrent, ix, x, xvi, 44, 202, 241-242, 245-253, 255-257, 259, 261, 264, 268, 281-282, 288, 291-293, 299-300, 309

Bram Cohen, 245, 249

Broadband, i, iii, xv, xvi, 3, 5, 30-34, 40, 47, 49, 67-68, 80, 97, 134, 137, 139, 178, 187-188, 197, 249, 281, 307-308, 311

C

Central processing unit (CPU), iii, 3-5, 16-20, 32, 56

Crucial 6

D

dBpowerAMP, iii, 89-90, 92

DDR, 6

Dial-up, xvi, 33, 187

Digital Audio Broadcasting (DAB), 23

Digital Rights Management (DRM), iv, 92, 116, 118-119, 133-137, 141, 143, 145-146, 181, 193, 212, 214, 281, 286-288, 307

Digital Subscriber Line (DSL), 31-33, 80

Direct Connect, 247

DivX iv, 146, 194-202, 206-210, 219, 225-230

DivX Converter, 199, 200, 201, 202

DMCA, 290-291

Downloading, ix, x, xi, xiv, xvii, 3, 9, 13, 25, 33-34, 40, 89, 134, 166, 169, 173, 220, 248, 256, 266, 274, 282, 301

DVD Decrypter, 199-200, 211-213, 221, 223, 235

DVD or CD?, 14

DVD Shrink, 209-210, 213-218, 234, 291

DVD X Copy, 290-291

E

ED2K, 261, 263, 267

eDonkey, xvi, 241, 246-248, 261, 268

eMule, ix, x, 202, 242, 246-248, 250, 261-263, 265-268

eMusic, iv, viii, 133, 135, 137, 168-173, 302

F

Fanning, Shaun, 243

File Sharing Networks, v, vii, ix, xi, xvi, 33, 39, 44, 177, 197, 202, 238-239, 241-243, 245-288, 293-304, 309

File transfer protocol (FTP), 69, 243

Filetopia, 246-247

Firewall, iii, 29-32, 39, 41, 43, 67-81, 100, 148, 242, 256-259, 267

Firewire, iii, 3-5, 9-13, 17, 29, 101

Flash memory, 15-16, 117

Framerate, 225, 233

Freeware, 46, 103, 129, 190, 211, 234

G

Gator, 55-56, 245

Gnutella, 241, 244, 247

Gordian Knot, 210, 218-221, 223-225, 227-230, 233

Graphics card, iii, 5, 16, 20, 22-23

Grokster, 244, 246, 293-296, 300

H

Hard drive, 3, 9, 15, 199, 207-208, 210

Hawkeye, 283

Hi-Fi, 15, 30, 88, 94, 106, 142

I

iFilm, 187-190

IFO, 196, 212, 223, 231, 235

iMix, 156-158

iTunes, see Apple iTunes

J

jetAudio, iv, 85, 92, 107-111, 147

K

Kad, 261, 267

Kazaa, 244-245, 276, 293, 296-297, 299-300

L

LaCie, 13, 208, 310

LAME, 92, 95

Index

Leechers, 249-251, 254, 260

LimeWire, 246, 248, 268, 276-277

LiteOn, 207-208

Live365, iv, 120-121

Malware, 25, 39, 42-44, 50-51, 53-54, 56-57, 66-67, 70, 247, 261, 268

Maxtor, 13

Merkur, 261

MicroATX, 17

MidiATX, 17

Morpheus, xvi, 244, 246-248, 268-272, 293, 300

Motherboard, iii, 3-6, 10, 16-17, 19-22, 26

MPEG-1, 92, 195, 197

MPEG-2, 23-24, 195, 197

MPEG-4, 195-196

MSN, iv, 116, 137, 138, 142-144, 163, 180-181, 283

Multimedia, xiii, xv, xvi, xvii, 3, 7, 11-14, 16, 20-22, 24-25, 28, 186, 191, 209, 290, 294-295, 308

Napster, iv, viii, xiv, xvi, 116, 118-119, 133, 136-137, 162-166, 168-169, 177-180, 243-244, 286-287, 289-290, 295

Open source, 25, 90, 92, 191, 196-197, 244

Peer-to-peer (P2P) 28, 162, 241-242, 259, 261, 264, 269, 293-294

PeerGuardian, 300

Playlist, 89, 102-103, 112-114, 141-142, 153, 156,-158, 160, 223

Plug-and-play, 9, 15, 118

Podcasts, 159-161

Pop-up, 55, 114, 122, 273

QuickTime, 146-147, 150, 188,-190, 196-197

R

RAR, 202, 204-205, 260

ratDVD, 209-210, 234,-237

Rewritable, 13-14

Ripping software, iii, 86, 90, 92, 102

S

SDRAM, 6

Secure Digital (SD), 15-16

Seeders, 250-251, 254-255, 258, 260

SlySoft, 193, 211

Solid-state discs, 15

Soulseek, 246-247

Sound card, iii, vii, 25-26, 124, 128

Spyware iii, 39, 54-57, 59, 64-66, 70, 82, 122, 190, 192, 241, 245, 248, 261, 268, 273, 276, 283

SpywareInfo, 246

Streaming, iii, iv, 5, 27-31, 34, 114-115, 119-123, 138, 144, 147, 181, 187-191, 195-196, 309-310

Supernode, 244, 269, 300

Swarm, 245, 251, 258

Sygate, 259

T

TFT versus CRT, 34

TorrentSpy, 252-257, 260, 309

TV tuner card, iii, 5, 23

V

VCDGear, 206

VideoLAN Media Player (VLC), 191, 194

VOB, 196-197, 222-223

VobSub, 220, 231, 233

vStrip, 223-224

W

Website, xi, 6, 33, 40, 43-44, 47-48, 53, 57, 59, 65, 89, 92, 107, 116, 124, 156, 175, 177, 179, 187, 190, 214, 243, 256, 299

Wi-Fi, 27-28, 30, 208, 308

WiMax, 308

WinAmp, iv, 112-115

Windows Media Player, iv, viii, 85, 96-99, 109, 112, 137-138, 140, 143, 146-147, 166, 174, 176, 185, 188, 191, 233-234

WinRAR, 202-205, 254

Wippit, viii, 133, 177-179, 283, 286, 302

WMV and MOV, 196

X

XP Plug and Play, 16

XviD, iv, 146, 195-198, 202, 207-208, 219, 225, 230, 238

Z

ZoneAlarm, 39, 41, 68, 70-73, 75-76, 78, 80-82, 259, 267